T0093033

MOLECULAR BIOLOGY INTELLIGENCE UNIT

VEGF in Development

Christiana Ruhrberg, PhD
Institute of Ophthalmology
University College London
London, UK

LANDES BIOSCIENCE
AUSTIN, TEXAS
USA

SPRINGER SCIENCE+BUSINESS MEDIA
NEW YORK, NEW YORK
USA

VEGF in Development

Molecular Biology Intelligence Unit

Landes Bioscience
Springer Science+Business Media, LLC

ISBN: 978-0-387-78631-5   Printed on acid-free paper.

Springer Science+Business Media, LLC, 233 Spring Street, New York, New York 10013, USA
http://www.springer.com

Please address all inquiries to the Publishers:
Landes Bioscience / Eurekah.com, 1002 West Avenue, 2nd Floor, Austin, Texas 78701, USA
Phone: 512/ 637 6050; FAX: 512/ 637 6079
http://www.landesbioscience.com

Printed in the United States of America.

9 8 7 6 5 4 3 2 1

Library of Congress Cataloging-in-Publication Data

VEGF in development / [edited by] Christiana Ruhrberg.
    p. ; cm. -- (Molecular biology intelligence unit)
  Includes bibliographical references and index.
  ISBN 978-0-387-78631-5 (alk. paper)
  1. Neovascularization. 2. Vascular endothelial growth factors. I. Ruhrberg,
Christiana. II. Series: Molecular biology intelligence unit (Unnumbered)
  [DNLM: 1. Vascular Endothelial Growth Factor A--physiology. 2. Growth and
Development--physiology. QU 107 V422 2008]
  QP106.6.V44 2008
  612.1'3--dc22
                                    2008009257

About the Editor...

CHRISTIANA RUHRBERG, PhD, is a Principal Investigator and Lecturer at University College London, working in the Institute of Ophthalmology. Having conducted her PhD thesis research in the laboratory of Dr. Fiona M. Watt at the Imperial Cancer Research Fund, Dr. Ruhrberg received a PhD in Biochemistry from Imperial College London. She subsequently obtained postdoctoral research experience in Dr. Robb Krumlauf's laboratory at the National Institute of Medical Research in London and in Dr. David Shima's laboratory at the Imperial Cancer Research Fund. Being funded largely by the UK Medical Research Council, her laboratory presently investigates the role of VEGF and neuropilins in vascular biology as well as neuronal and glial development.

=CONTENTS=

EDITOR

Christiana Ruhrberg
Institute of Ophthalmology
University College London
London, UK
Email: c.ruhrberg@ucl.ac.uk
Chapters 2 and 8

CONTRIBUTORS

Note: Email addresses are provided for the corresponding authors of each chapter.

Geert Carmeliet
Laboratory for Experimental Medicine
   and Endocrinology
K. U. Leuven
Leuven, Belgium
Chapter 7

Marcus Fruttiger
Institute of Ophthalmology
University College London
London, UK
Email: m.fruttiger@ucl.ac.uk
Chapter 3

Holger Gerhardt
Vascular Biology Laboratory
Cancer Research UK—London Research
   Institute
Lincoln's Inn Fields Laboratories
London, UK
Email: holger.gerhardt@cancer.org.uk
Chapter 6

Lauren C. Goldie
Department of Pediatrics
Children's Nutrition Research Center
and
Center for Cell and Gene Therapy
Baylor College of Medicine
Houston, Texas, USA
Chapter 4

Jody J. Haigh
Vascular Cell Biology Unit
Department for Molecular
   Biomedical Research
Flanders Interuniversity Institute
   for Biotechnology/Ghent University
Ghent, Belgium
Email: jody.haigh@dmbr.ugent.be
Chapter 5

Karen K. Hirschi
Department of Pediatrics
Children's Nutrition Research Center
and
Department of Molecular
   and Cellular Biology
Center for Cell and Gene Therapy
Baylor College of Medicine
Houston, Texas, USA
Email: khirschi@bcm.tmc.edu
Chapter 4

Janette M. Krum
Department of Anatomy
   and Cell Biology
The George Washington University
   Medical Center
Washington, DC, USA
Chapter 8

Christa Maes
Laboratory for Experimental Medicine
    and Endocrinology
K. U. Leuven
Leuven, Belgium
Email: christa.maes@med.kuleuven.be
Chapter 7

Yin-Shan Ng
(OSI) Eyetech
Lexington, Massachusetts, USA
Email: eric.ng@eyetech.com
Chapter 1

Melissa K. Nix
Department of Molecular
    and Cellular Biology
Center for Cell and Gene Therapy
Baylor College of Medicine
Houston, Texas, USA
Chapter 4

Jeffrey M. Rosenstein
Department of Anatomy
    and Cell Biology
The George Washington University
    Medical Center
Washington, DC, USA
Chapter 8

Quenten Schwarz
Institute of Ophthalmology
University College London
London, UK
Email: q.schwarz@ucl.ac.uk
Chapter 2

Joaquim Miguel Vieira
Institute of Ophthalmology
University College London
London, UK
Chapter 2

PREFACE

This book is devoted to vascular endothelial growth factor A (VEGF or VEGFA), a secreted signalling protein of great significance for development and disease in vertebrates. VEGFA controls the proliferation, migration, specialisation and survival of vascular endothelial cells, and it is therefore essential for the establishment of a functional blood vessel circuit. In addition, VEGFA is emerging as a versatile patterning factor for several non-endothelial cell types in vertebrates. Thus, it plays a central role during organ development at multiple levels, including blood vessel growth, vessel-mediated organ induction and tissue differentiation.

The discovery of non-endothelial roles for VEGFA initially came as a surprise. However, the recent indentification of a VEGF homolog in the fruit fly *Drosophila*[1,2] advocated the idea of an ancestral VEGF function in a non-endothelial cell type, as this organism lacks a vascular network. Most likely, VEGF's original role was to serve as a chemoattractant for various cell types, as VEGFA guides migrating haemocytes and border cells in *Drosophila*.[1,2] Because *Drosophila* VEGFA bears homology not only to VEGFA but also to the other VEGF superfamily members, including the platelet-derived growth factors (PDGFs) and placental growth factor (PGF), it was named VEGF/PDGF-related factor 1 (VPF1). It is now thought that the invertebrate VPF1 gene gave rise to the four vertebrate VEGF and PDGF genes during two rounds of genome duplication that were accompanied by functional mutation of the different VEGF gene copies.[2] These genome duplications also multiplied the gene for the VPF1 receptor, a tyrosine kinase named VPR1.[1] The VPR1 copies then evolved into the different vertebrate VEGF and PDGF receptors as well as several other receptor tyrosine kinases in the blood cell lineage.[2,3] In this fashion, several different ligand/receptor pairs became available to support the evolution of vertebrates with their increasingly complex and sophisticated organ systems. Most notably, the acquisition of a vascular circuit promoted intrauterine embryogenesis and provided the basis for increased organ and hence body size.

Two members of the VEGF superfamily in particular were recruited to support blood vessel formation, VEGFA to promote the growth of the inner, endothelial cell lining of blood vessels, and PDGFB to promote the acquisition of a supportive smooth muscle cell coat; in adults, PGF synergises with VEGF and PDGF to promote vascular repair.[4] Other VEGF homologs evolved to allow the elaboration of a lymphatic circuit (VEGFC/D)[5] or to guide the distribution of glial cell types in the ever-expanding central nervous system (PDGFA).[6] However, the boundaries that were initially drawn to help define the function of different VEGF superfamily members are becoming increasingly blurred; for example, VEGFC was recently reported to guide glial cells in the optic nerve,[7] whilst VEGFA was found to guide the migration of facial branchiomotor neurons in the brainstem.[8]

In writing this book, scientists investigating different aspects of VEGF signalling have come together to highlight the central importance of this growth factor for vertebrate development. From their reviews of our present knowledge, it becomes clear that VEGF's versatility as a patterning molecule is linked to its expression as several splice forms that interact with a number of differentially expressed signalling receptors. Moreover, by highlighting the present gaps in our knowledge, the authors of this book set direction to future research. It is therefore hoped that this book will provide valuable insights to those already studying VEGF as well as to those members of the scientific and medical communities that seek to understand vertebrate development and the origins of disease without specific prior knowledge of the field.

Christiana Ruhrberg, PhD

References

1. Duchek P, Somogyi K, Jekely G et al. Guidance of cell migration by the Drosophila PDGF/VEGF receptor. Cell 2001; 107(1):17-26.
2. Heino TI, Karpanen T, Wahlstrom G, et al. The Drosophila VEGF receptor homolog is expressed in hemocytes. Mech Dev 2001; 109(1):69-77.
3. Grassot J, Gouy M, Perriere G et al. Origin and molecular evolution of receptor tyrosine kinases with immunoglobulin-like domains. Mol Biol Evol 2006; 23(6):1232-1241.
4. Carmeliet P. Angiogenesis in health and disease. Nat Med 2003; 9(6):653-660.
5. Tammela T, Petrova TV, Alitalo K. Molecular lymphangiogenesis: new players. Trends Cell Biol 2005; 15(8):434-441.
6. Fruttiger M, Karlsson L, Hall AC et al. Defective oligodendrocyte development and severe hypomyelination in PDGF-A knockout mice. Development 1999; 126(3):457-467.
7. Le Bras B, Barallobre MJ, Homman-Ludiye J et al. VEGF-C is a trophic factor for neural progenitors in the vertebrate embryonic brain. Nat Neurosci 2006; 9(3):340-348.
8. Schwarz Q, Gu C, Fujisawa H et al. Vascular endothelial growth factor controls neuronal migration and cooperates with Sema3A to pattern distinct compartments of the facial nerve. Genes Dev 2004; 18(22):2822-2834.

Acknowledgments

I wish to thank all authors for sacrificing their valuable time to contribute chapters to this book.

CHAPTER 1

The Biology of Vascular Endothelial Cell Growth Factor Isoforms

Yin-Shan Ng*

Abstract

The field of angiogenesis research was literally transformed overnight by the discovery of vascular endothelial growth factor (VEGF). Researchers quickly embraced VEGF in their different areas of vascular and angiogenesis research, and in the last two decades have discovered much about VEGF biology. It is now clear that VEGF is actually a collection of different isoforms. Through differential pre-mRNA splicing and protein processing, one VEGF gene gives rise to several different protein isoforms, which together orchestrate the complex processes of angiogenesis, vessel growth and adult vascular functions. The VEGF isoforms differ biochemically, and genetic experiments in mice have proven that the isoforms have different functions. Furthermore, certain VEGF isoforms associate with and likely play differential roles in various pathologic states. With better understanding of VEGF isoform biology, new insights into the complex mechanisms of VEGF-mediated vessel growth can be gained. In addition, findings about the specific VEGF isoform functions have important implications for VEGF-mediated therapeutic angiogenesis as well as anti-angiogenic therapy targeting VEGF.

Key Messages
- VEGF is a collection of different isoforms
- VEGF isoforms have different biochemical properties
- VEGF isoforms have overlapping and distinct functions during development
- VEGF is multifunctional for endothelial cells and also acts on other cell types
- Expression of a specific VEGF isoform is associated with pathological conditions

Introduction

Angiogenesis plays a critical role in the progression of many pathologies, including cancer. The search for angiogenic factors in the last decade has been largely driven by the hope that identification of such factors will lead to new treatments for these pathologies. Anti-angiogenic therapy is particularly promising for the treatment of cancer; the strategy of blocking tumor angiogenesis seems to offer the best approach yet for treating tumors resistant to conventional therapies.[1] The list of putative angiogenesis factors grows continuously, but VEGF, one of the first angiogenesis factors identified, is widely believed to be the most important regulator of both normal and pathological angiogenesis.

VEGF was first purified from tumor cells by Harold Dvorak and coworkers at the Beth Israel Hospital in Boston, USA. This factor was isolated based on its ability to enhance vascular

*Yin-Shan Ng, (OSI) Eyetech, 35 Hartwell Ave, Lexington, Massachusetts 02421, USA.
Email: eric.ng@eyetech.com

VEGF in Development, edited by Christiana Ruhrberg. ©2008 Landes Bioscience and Springer Science+Business Media.

permeability, and was therefore named "vascular permeability factor" or VPF.[2] Napoleon Ferrara at Genentech subsequently purified, and later cloned, a factor from medium conditioned by bovine pituitary folliculostellate cells that induced proliferation of vascular endothelial cells (EC). This substance was named VEGF.[3] When cDNAs from both the Dvorak and the Ferrara studies were sequenced, VEGF and VPF were found to be the same molecule.

It is now clear that VEGF elicits an array of EC responses in vitro, including stimulation of proliferation and migration, and induction of proinflammatory gene expression.[4] VEGF has been shown to guide blood cell and EC precursor migration in *Drosophila*,[5] *Xenopus*,[6] and likely in *Danio rerio*;[7] it is possible that in mammals the ancestral role of VEGF is to direct EC precursor, or angioblast, migration.[8] The findings in the *Drosophilia* and in the lower vertebrates suggest that at least the chemotaxic function of VEGF signaling is conserved during evolution, and the EC mitogenic and vascular permeability functions of VEGF may have evolved in more complex animals to modulate their vascular system.

Certain in vivo characteristics of VEGF further illustrate its role as the primary angiogenic factor. VEGF is a secreted angiogenic factor, and thus can act in a paracrine fashion.[4] Expression patterns of VEGF and its endothelial-selective receptors correlate both temporally and spatially with areas of vascular growth in developing embryos and during the female reproductive cycle.[4,9,10] Most importantly, embryos lacking either VEGF or its receptors fail to develop functional vessels.[11-14] Interestingly, heterozygous VEGF gene inactivation also results in severe disruption of vessel development and embryonic lethality, suggesting that correct dosage of VEGF is critical for normal vascular development. Indeed, further genetic experiments using a hypermorphic VEGF allele confirmed the tight dosage requirement of VEGF in cardiovascular development.[15] Lastly, antagonists of VEGF or its receptors can effectively block both normal and pathological angiogenesis in animals.[16-18]

In addition to its role in developmental angiogenesis, VEGF also modulates adult physiological angiogenesis and vessel function in numerous pathologies. In the adult, VEGF participates in regulation of the female reproductive cycle, wound healing, inflammation, vascular permeability, vascular tone, and hematopoiesis.[19] VEGF function also contributes to pathological angiogenesis in disorders such as cancer, rheumatoid arthritis, diabetic retinopathy and the neovascular form of macular degeneration.[4,20,21]

VEGF, also called VEGFA, belongs to the cystine-knot superfamily of growth factors that are characterized by the presence of eight conserved cysteine residues.[22,23] In addition to VEGFA, this superfamily of hormones and growth factors includes VEGFB, VEGFC, VEGFD, VEGFE encoded by the various *orf* viruses, and placental growth factor (PlGF1 and PlGF2 isoforms).[22] The VEGF family members are active as secreted, glycosylated homodimers.[23] They are closely related to and likely have a common ancestor with the platelet derived growth factor (PDGF) family of growth factors, which includes PDGFB, PDGFC, and PDGFD.[22] VEGF exerts its effects on EC through binding to the EC-selective high affinity receptors FLT1 (VEGFR1), KDR (VEGFR2) and neuropilins (NRP1 and NRP2).[4,24] VEGF is highly conserved across different vertebrate species, from mammals to fish, at both the polypeptide and genomic structure levels. VEGF protein from one species is therefore often functional toward EC from another species.

VEGF Isoform Structure

The VEGF gene is comprised of several exons separated by introns in all species characterized—from human to mouse, fish and frog.[22,25] As a result of differential pre-mRNA splicing, a single VEGF gene gives rise to many different VEGF isoforms.[23] Because the roles of the different VEGF isoforms in nonmammalian species have not been well characterized, and most studies using specific VEGF isoforms have utilized either mouse or human models, this chapter will focus on the human and murine VEGF isoforms.

The gene encoding VEGF is located on the short arm of chromosome 6 (6p21.1) in humans and on chromosome 17 (24.20 cM) in the mouse.[22] Both the human and the murine VEGF genes are comprised of eight exons, separated by seven introns (Fig. 1). Although in theory any combination of the eight exons is possible, all differentially spliced variants of VEGF isoforms discovered to date contain the first 5 exons (1 to 5) plus different combinations of exons 6 to 8. As a result, all VEGF isoforms contain signal peptides (first 28 residues) and are thus secreted polypeptides, and they can all potentially form homodimers, because the dimerization domain is located in exons 2 to 5 (residues Cys51 and Cys61).[23] Because three residues tially spliced region of the VEGF pre-mRNA is limited to exons 6 to 8, the number of potential possible combinations of these exons is further increased by the fact that exons 6 and 7 have alternative internal splice donor and acceptor sites that further divide them into two different portions, referred to as 6a and 6b, 7a and 7b, respectively.[22,25]

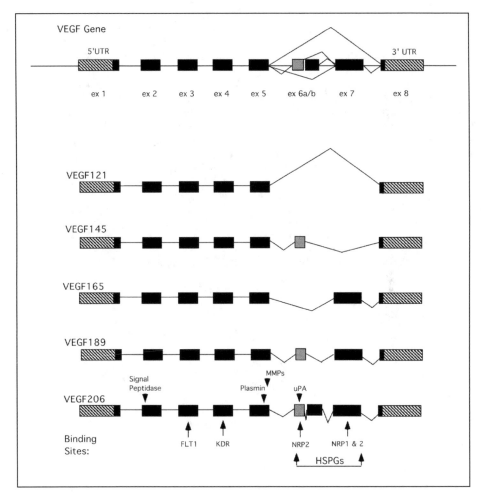

Figure 1. Genomic organization of the human VEGF gene and its alternative splice variants. By differential pre-mRNA splicing of exons 6 and 7, the human *vegf* gene gives rise to various isoforms (the most common 5 isoforms are shown). The different exons (ex) and domain sizes (a.a., amino acid residue) are shown below the gene. The 5'- and 3'-untranslated regions (UTR) are indicated by hatched boxes. Protease cleavage sites (arrowheads) and receptor binding sites (arrows) are indicated on the VEGF206 isoform.

Despite this variety of potential differential splicing combinations, it appears that there are only three major VEGF isoforms produced in all vertebrates, and these differ by the presence or absence of peptides encoded by exons 6 (24 amino acids) and 7 (44 amino acids). The major human VEGF isoforms are VEGF121, VEGF165 and VEGF189, with the numbers indicating the number of amino acids in the mature polypeptides (Fig. 1). The major murine VEGF isoforms are VEGF120, VEGF164 and VEGF188,[26] each containing one less amino acid than the human orthologue. In vivo, VEGF121(120), VEGF165(164) and VEGF189(188) isoforms are produced by certain cells in a tissue-specific pattern with the clear exception of most vascular endothelial cells,[27] and many of the less-abundant VEGF isoforms are associated with specialized cell types or tumor cells (see below).

Biochemistry of the Major VEGF Isoforms

The different VEGF isoforms have distinct biochemical properties (Fig. 2). As VEGF120(121) does not bind heparan sulfate, it is readily diffusible. VEGF164(165) has moderate affinity for heparan sulfate; it is partially sequestered on the cell surface and in the extracellular matrix (ECM),[28] likely due to heparan sulfate proteoglycans (HSPGs) binding. The HSPGs-binding activity of VEGF164(165) is conferred by the 15 basic residues within the peptide encoded by exon 7, which is defined as the heparin-binding domain of VEGF.[23] VEGF188(189) has high affinity for heparan sulfate due to the presence of the additional basic residues and a strongly heparin-binding domain encoded by exon 6. As a result, VEGF188(189) is mostly associated with the cell-surface and ECM.[28] (In this chapter, the nomenclature for the human and murine VEGF isoforms will be used interchangeably).

The VEGF isoforms localized or sequestered at the cell surface or ECM constitute a reservoir of angiogenic growth factors that can be mobilized by various enzymes. For example, heparinase and matrix metalloproteinases (MMPs) can release matrix-bound VEGF isoforms from HSPGs,[23,28] whereas plasmin can cleave the heparin-binding domain of the matrix-bound

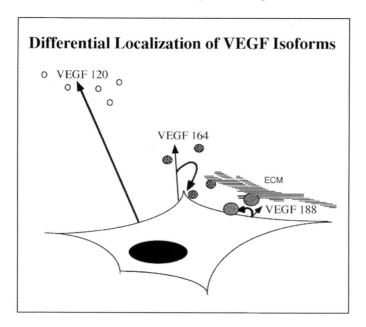

Figure 2. Distinct biochemistry of the different VEGF isoforms. Schematic representation of the differential extracellular localization of the three murine VEGF isoforms based on their different affinities for heparin sulfate.

VEGF isoforms, producing a truncated, but biologically active, VEGF110 (VEGF109 in the mouse).[28] This enzyme-mediated VEGF release represents a fast and effective way to mobilize VEGF and increase its effective concentration in different local environments. Therefore one of the functions for the different VEGF isoforms is to modulate the availability of VEGF by being sequestering on the cell surface or in the ECM. Although the details of this protease-mediated VEGF release pathway have not been elucidated in vivo, MMP9 plays a major role in VEGF-mediated tumor angiogenesis in the RipTag pancreatic tumor model[29] and during bone development,[30] suggesting that protease-mediated VEGF release does indeed play a prominent role in the VEGF pathway. In addition to releasing matrix-sequestered VEGF, it has been reported that a nMMP with anti-angiogenic properties, ADAMTS1, directly binds to VEGF and dampens downstream VEGF signaling.[31] More recently, it has been reported that many different MMPs can cleave the matrix-bound VEGF isoforms intramolecularly, creating two fragments: a soluble and biologically active N-terminal fragment that can bind to the VEGF receptors, and a shorter matrix-binding C-terminal fragment that has no apparent biological activity.[32] Interestingly, the authors also reported distinct angiogenic outcomes produced by the matrix-bound VEGF and the MMP-processed soluble N-terminal VEGF fragment, providing further proof that matrix association can modulate the function of the different VEGF isoforms. These findings clearly suggest that interactions of various MMPs and possibly other proteases with VEGF occur in vivo, and that these interactions can directly modulate both the availability and the activity of VEGF.

Not only do the VEGF isoforms display differences in localization and availability, they also have different affinities for their high affinity receptors, FLT1 and KDR, expressed on the EC surface. Binding of FLT1 by both VEGF121 and VEGF165 is inhibited by heparin. Binding of KDR by VEGF165 is enhanced by low concentrations of heparin, but is inhibited by high concentrations of heparin.[23] Binding of KDR by VEGF121 is not affected by heparin. Besides differential binding affinities for FLT1 and KDR, the neuropilin family of cell-surface receptors exhibits differential specificity for VEGF isoforms. Because NRP1 binds to the peptide encoded by exon 7, VEGF165 and probably VEGF189 bind to NRP1, but VEGF121 does not.[24] Although binding by VEGF165 may not induce NRP1 signal transduction directly, due to the lack of a conventional signaling cytoplasmic domain, it was reported that cell surface NRP1 can increase the binding affinity of VEGF165 for KDR,[24,33] but not for VEGF121. This suggests that NRP1 functions as a coreceptor in EC cells. Interestingly, it has been reported that VEGF165, but not VEGF121, can promote survival of breast carcinoma cells in vitro in a neuropilin-dependent manner, suggesting that NRP1 may function as a signaling VEGF165-specific receptor in nonendothelial cells.[34] Indeed, a chimeric receptor containing the transmembrane and intracellular domain of NRP1 and the extracellular domain of epidermal growth factor (EGF) was reported to mediate human umbilical vein EC (HUVEC) migration upon EGF stimulation.[35] These results suggest that the NRP1 receptor is capable of transducing signals in certain cell types and in certain contexts. Considering the important role of the NRP1 receptor reported for vascular development[23,36] and for guidance of EC tip cells and vessels,[37] more studies into the exact signaling role of the NRP1 will provide further insights into the biology of VEGF isoforms and their roles in modulating angiogenesis. Another neuropilin family member, NRP2, has also been shown to bind VEGF165 but not VEGF121.[38] Interestingly, NRP2 was reported to function as a receptor for the less-abundant VEGF145 isoform,[38] suggesting that both exon 6 and exon 7 of VEGF can facilitate the binding to NRP2 by VEGF145 and VEGF165, respectively. However, the functional role of this selective interaction between VEGF isoforms and NRP2 in angiogenesis remains to be determined.

It is likely that the difference in biochemical properties described above translates into distinct biological activities for the various VEGF isoforms. For example, it has been reported that the VEGF isoforms can have different mitogenic activity for EC in vitro and in vivo.[23] However, others have found that VEGF120 and VEGF164 do not differ in their ability to support EC proliferation, but that vessel networks differ in tissues developing in the absence of specific isoforms and in tumors overexpressing the various VEGF isoforms in vivo.[32,39-43] The differences in localization, availability, receptor-binding affinity and bioactivity likely contribute to distinct roles for the individual VEGF isoforms during vascular development, and imply that the presence of the different VEGF isoforms is critical for normal vascular development. Therefore, knowledge of the differential functions of the different isoforms will be crucial for designing an effective VEGF-mediated angiogenesis therapy to promote normal vessels growth in patients suffering from ischemic vascular diseases.[44]

VEGF Isoforms in Vascular Development

VEGF mediates angiogenesis both during development and in the adult. While the roles of specific VEGF isoforms in normal adult angiogenesis remain largely unexplored, contributions of VEGF isoforms to developmental angiogenesis have been identified using transgenic mouse models.

Experiments that directly measured the VEGF isoform mRNA levels in the mouse revealed that the relative levels of the three major VEGF isoforms vary among different adult organs. Differential mRNA levels of the three major VEGF isoforms during murine embryonic development also suggest that expression of the different isoforms is developmentally regulated. Whereas the three major VEGF isoforms are expressed in all embryonic organs examined, the relative levels of each isoform varied from organ to organ, and the isoform composition changed over developmental time in the same organ.[26] For example, adult lung, heart and liver express relatively high levels of VEGF188 mRNA (52% of the total VEGF message in the lung and 36% in the heart and liver), whereas brain, eye, spleen and kidney express relatively low levels of VEGF188 mRNA (6%, 5%, 11% and 17% of the total VEGF mRNA, respectively). During development, in the embryonic day 13.5 lung, VEGF188 makes up only about 10% of total VEGF mRNA, but at embryonic day 17.5, just before birth, about 50% of the total VEGF mRNA produced in the lung is the VEGF188 isoform. The levels of VEGF188 remain high at about 50% of total VEGF mRNA in the adult lung. These findings are consistent with the concept that the VEGF isoforms serve different functions during vascular development and angiogenesis in the adult.[26]

To directly assess the different functions of the three major VEGF isoforms during development and in the adult, and in an attempt to avoid the early embryonic lethality from the total gene inactivation,[13,14] VEGF isoform-specific gene targeting approaches were used. Using exon-specific deletion and cDNA replacement (knock-in) strategies, VEGF alleles were created that permitted expression of only one of the three VEGF isoforms: VEGF120, VEGF164 or VEGF188.[39,40] Mice were then generated in which the normal VEGF gene was replaced with one or two isoform-specific alleles. The isoform-specific mouse models yielded two immediate and significant findings. First, heterozygous mice in which one of the two VEGF alleles could produce only a single VEGF isoform were viable and developed normally (Table 1). Given that single-allele VEGF inactivation, as well as the slight increase in VEGF levels resulting from a hypermorphic mutation, both lead to embryonic lethality,[15] these observations suggest that the isoform-specific allele likely produced similar levels of VEGF transcript compared to the wild-type allele. Indeed, RNase protection assays confirmed that the total levels of VEGF transcript in the heterozygous and the homozygous VEGF isoform-specific mice were comparable to that of wild-type mice.[39] A second important finding was that although not all VEGF isoform-specific homozygous embryos were viable (Table 1), their phenotypes were much less severe than those of VEGF-null mice. This result is consistent with the concept that the different VEGF isoforms all support EC growth, but have subtle differences in function, such that one isoform can only partially replace the function of the others during embryonic development and in the adult (Table I).

Analysis of the VEGF[120/120] and VEGF[188/188] homozygous mice, in which both VEGF alleles produce only a single VEGF isoform (VEGF120 and VEGF188, respectively), has yielded further insight into the roles of the VEGF isoforms in vascular development. Homozygous VEGF[120/120] mice are not viable and exhibit very distinct vascular defects. Most of the VEGF[120/120] embryos die in utero, or soon after birth. All VEGF[120/120] embryos exhibit decreased angiogenesis,

Table 1. The phenotypes of VEGF isoform-specific mice

Homozygous	**vegf120/120** Perinatal (90%) or early postnatal (10%) lethality. Extensive vascular hemorrhages in most organs. Decreased vascular branching and impaired postnatal angiogenesis[26,39,43] (Fig. 3).	**vegf164/164** Viable, normal appearance[40] (P. D'Amore and Y. S. Ng, unpublished data). No obvious vascular phenotypes in the embryo (C. Ruhrberg and D.T. Shima, unpublished data). Yolk sac vesselsare patterned normally (Fig. 3).	**vegf188/188** Embryonic lethality from embryonic day 10 (90%); 10% reach adulthood, though with decreased body size/weight and subtle vascular defects in the retina[40] (P. D'Amore and Y. S. Ng, unpublished data). Excessive branching and small caliber of vessels in several embryonic organs.[43] Hyper-fused vessel networks in the yolk sac (Fig. 3).
Heterozygous	**vegf120/+** Viable, normal appearance, but decreased vascular branching in several embryonic organs[26,39,43] (Fig. 3).	**vegf164/+** Viable, normal appearance, normal vessel patterning in all organs examined[40] (P. D'Amore and Y. S. Ng, unpublished data; C. Ruhrberg and D. T. Shima, unpublished data).	**vegf188/+** Viable, normal appearance[40] (P. D'Amore and Y. S. Ng, unpublished data). Vascular phenotype not examined.
Compound heterozygous	**vegf164/120** Viable, normal (P. D'Amore and Y. S. Ng, unpublished data). Viability suggests that the combination of VEGF164 and VEGF120 isoforms is sufficient for normal vascular development.	**vegf188/120** Viable, normal (P. D'Amore and Y. S. Ng, unpublished data). Normal vascular development in several organs.[43,46] These observations suggest that the combination of VEGF188 and VEGF120 isoforms is sufficient for normal vascular development.	**vegf188/164** Viable, normal (P. D'Amore and Y. S. Ng, unpublished data). The vascular phenotype has not been examined. Viability suggests that the combination of VEGF188 and VEGF164 isoforms is sufficient for normal vascular development.

Mice expressing different combinations of VEGF isoforms were produced by crossing heterozygous mice carrying one wild-type and one isoform-specific *vegf* allele.

decreased vascular branching and vascular hemorrhage,[39] suggesting that the vessels in these mice are leaky and lack the normal vascular morphology found in wild-type mice. Close examination of the vascular beds in organs from the VEGF$^{120/120}$ mice of different developmental stages revealed extensive vascular remodeling defects and decreased vascular branching and density (Fig. 3).[26,43] These results suggest that the freely diffusible VEGF120 isoform alone can support the initial stages of vascular development, but that VEGF120 cannot replace the functions of the heparin-binding VEGF isoforms such as VEGF164 in the fine patterning of the vasculature. There is also a strong correlation between the severity of vascular phenotypes observed in the VEGF$^{120/120}$ embryos and the organ expression patterns of the various isoforms; that is, the organs that express relatively high levels of the heparin-binding VEGF isoforms (VEGF164 and VEGF188) exhibited the most severe, abnormal vascular phenotypes. For example, high levels of

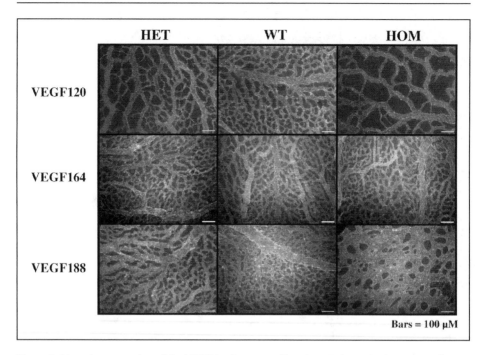

Figure 3. Vascular patterning of the VEGF isoform-specific mice. Embryonic day 10.5 yolk sacs showing the different vascular patterns formed in the VEGF120, VEGF164, and VEGF188 isoform-specific mice. In the heterozygous (HET) and homozygous (HOM) VEGF120 yolk sacs, the lack of vessel sprouting results in sparser vascular networks as compared to wildtype (WT), whereas the HOM VEGF188 yolk sac displayed enlarged and sheet-like vessels likely resulting from fusion of vessels due to excessive vascular sprouting and/or branching. Vessels in the VEGF164 yolk sacs are indistinguishable from the WT.

VEGF188 expression are temporally and spatially associated with the maturation of the lung, and analysis of the lungs from $VEGF^{120/120}$ mice revealed extensive microvascular defects and retarded alveolarization. Thus, although the VEGF120 isoform alone appears to support initial vessel growth in the lung, remodeling or maturation/maintenance of the final vascular bed, including further angiogenesis, is defective in the absence of the heparin-binding VEGF isoforms.[26]

Because the $VEGF^{120/120}$ mice produce only freely soluble VEGF, defects in these mice provide information about potential roles of heparin-binding isoforms such as VEGF164. Analysis of organs in $VEGF^{120/120}$ mice that are vascularized by angiogenic sprouting, such as the hindbrain suggested that matrix-bound VEGF164 is required to establish a normal concentration gradient of VEGF. This concentration gradient, formed when VEGF164 binds to the cell surface and ECM, can effectively direct the chemotactic activity of the EC by modulating the behavior of filopodia on endothelial vessel tips during vessel sprouting[43] (see Chapter 6 by H. Gerhardt). As the VEGF120 isoform is freely diffusible, these gradients cannot form properly in $VEGF^{120/120}$ mice. These observations establish the importance of localized VEGF in establishing a concentration gradient, and help to explain the unique role of the heparin-binding function of VEGF isoforms in vascular development.

VEGF$^{188/188}$ mice, too, display abnormal vascular development. About half of the homozygous VEGF$^{188/188}$ embryos die around embryonic day 10.5, exhibiting developmental delay and extensive vascular defects (D'Amore and Ng unpublished data), and the remaining half develop to term with subtle vascular defects in most organs examined.[40] Because the VEGF188 isoform is highly localized to ECM and cell surface HSPGs, proteolytic cleavage is required to release the biologically active form of soluble N-terminal VEGF fragment from the cell surface and ECM in tissue culture experiments.[23,28,32] The fact that some VEGF[188/188] mice are viable suggests that the VEGF188 isoform is processed into soluble receptor-binding N-terminal fragment and in ECM-binding C-terminal fragments by proteases in vivo. The VEGF[188/188] mouse model supports the idea that post-translational proteolytic processing of VEGF isoforms in vivo could potentially create additional isoform proteins beyond those formed by differential pre-mRNA splicing. However, the functional significance of post-translational modification/processing of VEGF isoforms remains unclear, and more studies will be required to elucidate the biology of this pathway in vascular development and neoangiogenesis.

VEGF Isoforms in Diseases

A theory is emerging that particular VEGF isoforms might associate with certain disease states;[45] in this case, specific VEGF isoforms might be considered pathological. Although the mechanism of the disease/pathological VEGF isoform association is unclear, understanding of isoform specificities for diseases might permit development of better and more specific anti-VEGF therapies to treat these conditions.

Amongst the better-defined VEGF-mediated pathologies are the proliferative intraocular neovascular syndromes in diseases such as diabetic retinopathy, retinopathy of prematurity (ROP), and the wet form of macular degeneration. In all of these disease states, VEGF not only causes uncontrolled neovascular growth that damages the retina, but also promotes vascular leakage and vitreous hemorrhages that eventually lead to blindness.[21] Recently, VEGF164 has been identified as the major pathological VEGF isoform in the eye.[21,46] In experimental murine models of diabetes and ROP, VEGF164 was more potent than VEGF120 in inducing both endothelial intercellular adhesion molecule 1 (ICAM-1) expression and chemotaxis of leukocytes, which together lead to increased inflammation of the retinal vessels. Furthermore, VEGF164 was more potent than VEGF120 in inducing vascular leakage and blood-retinal barrier breakdown. Interestingly, the vascular inflammation and leakage caused by VEGF in these models were significantly reduced following administration of a pan-VEGF isoform antagonist or a VEGF164-specific, RNA-based aptamer antagonist.[21] These observations suggest that VEGF164 is the major disease-causing isoform in models of neovascular eye disease, and highlight the importance of understanding the different contributions of specific VEGF isoforms in vascular pathologies. Therefore, anti-angiogenic therapy targeted to individual VEGF isoforms might increase the specificity and potentially the efficacy of the therapy.

Another disease that is dependent on angiogenesis is cancer, as solid tumors incorporate new vessels to support and promote tumor growth and metastatic spread.[1] Although not all tumor angiogenesis is VEGF dependent, most tumor types studied to date display upregulation of VEGF mRNA and protein.[4] It is therefore likely that VEGF is at least partly responsible for tumor angiogenesis in most cancer types. Interestingly, tumor vessels are mostly unstable, leaky and immature; these characteristics result from the high levels of VEGF in the tumor.[47] VEGF164 and VEGF120 appear to be the most widely expressed VEGF isoforms in tumors. One early study in which murine brain tumors over-expressed one of the three major VEGF isoforms, showed that both VEGF120 and VEGF164 promoted rapid growth of vessels that were highly unstable and leaky. In contrast, the VEGF188 over-expressing brain tumor supported vessels that were slower growing, nonhemorrhagic vessels that were relatively normal in appearance.[41] These results demonstrated that all three VEGF isoforms can stimulate angiogenesis, but that the characteristics of the resulting vessels depend on the specific isoform(s) expressed.

Because this tumor study was complicated by the fact that VEGF was also endogenously expressed by the tumor cells, a more refined study was conducted in which tumor cells derived from VEGF-deficient fibroblasts were used to create VEGF isoform-specific tumor cells.[42] In the mouse, these VEGF isoform-specific cells all supported vessel growth in the fibrosarcomas, but the new tumor vessels displayed different vascular characteristics depending on the expressed isoform. The VEGF164-expressing cells induced tumor growth and vessel density similar to those observed for wild-type tumor cells. The VEGF120-specific cells only partially rescued tumor growth, and vascular density in these tumors was reduced. Although the VEGF188-specific cells formed smaller tumors than wild type cells, the vessels in these tumors were present at a higher density and displayed a more highly branched phenotype than observed for the wild type tumor. Furthermore, by mixing the VEGF120- and VEGF188-specific cells, both the vascular density and tumor growth were comparable to those in the wild type tumor. Taken together, these results suggest that different VEGF isoforms play distinct but cooperative roles during tumor angiogenesis. Although VEGF164 and VEGF120 were more potent than VEGF188 in inducing both tumor growth and angiogenesis, the isoform-specific fibrosarcomas also demonstrated that the VEGF isoforms could act cooperatively to make normal-appearing and functional vessels during tumor vascularization.[42]

Specific VEGF isoform expression has also been associated with the progression of a particular tumor type in humans,[48] and may be an indicator for the prognosis of the disease. For example, high levels of VEGF189 expression were associated with nonsmall-cell lung cancer (NSCLC) and were an indicator of poor prognosis in patients, compared to NSCLC expressing low levels of VEGF189. High levels of VEGF165 expression were associated with poor prognosis in osteosarcoma.[48] Although it is not clear whether the predominance of a particular VEGF isoform in a tumor is the cause or a consequence of the disease, such associations might be used as disease markers with potential diagnostic or therapeutic value.[45]

Novel VEGF Isoforms

With the improved sensitivity of RT-PCR detection techniques and better design of PCR primers based on the VEGF genomic DNA sequence, many novel and low-expression VEGF isoforms have been discovered in diverse tissue types. Relatively low levels of VEGF206 (exons 1-5, 6a, 6b, 7 and 8) and VEGF145 (exons 1-5, 6a, 8) transcripts have been detected in normal human placental tissues, and VEGF145 was the major isoform expressed in several carcinoma cell lines derived from the female reproductive system.[23] VEGF183 (exons 1-5, truncated 6a, 7 and 8) is similar to VEGF189 but with a shorter exon 6a, and has been reported to have wide tissue distribution.[23] A new heparin-binding VEGF162 (exon 1-5, 6a, 6b, 8) has recently been discovered in human ovarian carcinoma cells, and has been reported to have angiogenic activity.[49] VEGF165b (exon 1-5, 7, distal splice site of exon 8) is very similar to VEGF165 in protein sequence and size, but inhibits VEGF165 by competing for KDR binding,[50] and thus may function as an endogenous, competitive inhibitor of all VEGF isoforms for receptor binding.

Pre-mRNA splicing is very important in normal development, as it creates protein diversity in complex organisms. This mechanism is also a natural target for various disease processes, including carcinogenesis. Since mice expressing only VEGF164 are viable,[40] the novel isoforms may be unnecessary for the normal development and survival of adults (Table 1). These novel isoforms may, however, play important roles in disease states. Because upregulation of novel VEGF isoforms has been observed predominately with abnormal tissues, it will be important to determine if these isoforms play a role in disease progression. In vivo animal models will be needed to correlate and confirm expression of novel VEGF isoforms, to exclude potential tissue culture artifacts. Further studies of these isoforms, with their different combinations of C-terminal exons, might help to better elucidate the functions of VEGF in vivo.

Conclusions

The central role of VEGF in both normal and pathological angiogenesis has been established, and the complex biology of VEGF, including roles of the different isoforms and receptors, is starting to be elucidated. From the vascular biology point of view, however, much remains unclear regarding the differential functions of the VEGF isoforms. For example, do specific pairings of isoforms with VEGF receptors contribute to the different functions of the VEGF isoforms? Do the isoforms have differential effects on vascular permeability, which is a major problem associated with many vascular pathologies? VEGF was once considered an endothelial-specific growth factor that mediating angiogenesis and permeability, but it is now clear that VEGF has additional functions beyond the vasculature that affect cell types including, for example, neural cells, bone-forming cells and immune cells (see Chapters 8 by J. Haigh, J.M. Krum and C. Ruhrberg; Chapter 7 by C. Maes and G. Carmeliet; and Chapter 3 by J.J. Haigh). The roles of the VEGF isoforms in mediating inflammation are poorly understood. Furthermore, the differential effects of the VEGF isoforms on the nervous system, likely via the NRP1/2 receptors, in development and in diseases remain largely unexplored. With the increasing interest in the use of anti-VEGF therapy to treat various diseases that are associated with abnormal angiogenesis, inflammation, and vessel hyperpermeability, more attention must be paid to understanding the exact roles of VEGF isoforms in normal physiology and vascular pathologies. A better understanding of the isoform functions in building and maintaining normal vessels during development will also allow the design of better pro-angiogenic therapies; specific VEGF isoforms might be particularly beneficial for patients suffering from ischemic diseases.

Acknowledgements

The author would like to thank Patricia A. D'Amore for providing Fig. 3 and Anne Goodwin for critical reading of the manuscript.

References

1. Folkman J. Role of angiogenesis in tumor growth and metastasis. Semin Oncol 2002; 29(6 Suppl 16):15-18.
2. Senger DR, Connolly DT, Van de Water L et al. Purification and NH2-terminal amino acid sequence of guinea pig tumor-secreted vascular permeability factor. Cancer Res 1990; 50(6):1774-1778.
3. Ferrara N, Henzel WJ. Pituitary follicular cells secrete a novel heparin-binding growth factor specific for vascular endothelial cells. Biochem Biophys Res Commun 1989; 161(2):851-858.
4. Ferrara N. Vascular endothelial growth factor: Basic science and clinical progress. Endocr Rev 2004; 25(4):581-611.
5. Cho NK, Keyes L, Johnson E et al. Developmental control of blood cell migration by the Drosophila VEGF pathway. Cell 2002; 108(6):865-876.
6. Cleaver O, Krieg PA. VEGF mediates angioblast migration during development of the dorsal aorta in Xenopus. Development 1998; 125(19):3905-3914.
7. Liang D, Chang JR, Chin AJ et al. The role of vascular endothelial growth factor (VEGF) in vasculogenesis, angiogenesis, and hematopoiesis in zebrafish development. Mech Dev 2001; 108(1-2):29-43.
8. Traver D, Zon LI. Walking the walk: Migration and other common themes in blood and vascular development. Cell 2002; 108(6):731-734.
9. Breier G, Albrecht U, Sterrer S et al. Expression of vascular endothelial growth factor during embryonic angiogenesis and endothelial cell differentiation. Development 1992; 114(2):521-532.
10. Millauer B, Wizigmann-Voos S, Schnurch H et al. High affinity VEGF binding and developmental expression suggest Flk-1 as a major regulator of vasculogenesis and angiogenesis. Cell 1993; 72(6):835-846.
11. Fong GH, Rossant J, Gertsenstein M et al. Role of the Flt-1 receptor tyrosine kinase in regulating the assembly of vascular endothelium. Nature 1995; 376(6535):66-70.
12. Shalaby F, Rossant J, Yamaguchi TP et al. Failure of blood-island formation and vasculogenesis in Flk-1-deficient mice. Nature 1995; 376(6535):62-66.
13. Carmeliet P, Ferreira V, Breier G et al. Abnormal blood vessel development and lethality in embryos lacking a single VEGF allele. Nature 1996; 380(6573):435-439.
14. Ferrara N, Carver-Moore K, Chen H et al. Heterozygous embryonic lethality induced by targeted inactivation of the VEGF gene. Nature 1996; 380(6573):439-442.
15. Miquerol L, Langille BL, Nagy A. Embryonic development is disrupted by modest increases in vascular endothelial growth factor gene expression. Development 2000; 127(18):3941-3946.
16. Kim KJ, Li B, Winer J et al. Inhibition of vascular endothelial growth factor-induced angiogenesis suppresses tumour growth in vivo. Nature 1993; 362(6423):841-844.
17. Millauer B, Shawver LK, Plate KH et al. Glioblastoma growth inhibited in vivo by a dominant-negative Flk-1 mutant. Nature 1994; 367(6463):576-579.

18. Adamis AP, Shima DT, Tolentino MJ et al. Inhibition of vascular endothelial growth factor prevents retinal ischemia-associated iris neovascularization in a nonhuman primate. Arch Ophthalmol 1996; 114(1):66-71.
19. Ferrara N. Role of vascular endothelial growth factor in regulation of physiological angiogenesis. Am J Physiol Cell Physiol 2001; 280(6):C1358-1366.
20. Folkman J. Angiogenesis in cancer, vascular, rheumatoid and other disease. Nat Med 1995; 1(1):27-31.
21. Adamis AP, Shima DT. The role of vascular endothelial growth factor in ocular health and disease. Retina 2005; 25(2):111-118.
22. Holmes DI, Zachary I. The vascular endothelial growth factor (VEGF) family: Angiogenic factors in health and disease. Genome Biol 2005; 6(2):209.
23. Robinson CJ, Stringer SE. The splice variants of vascular endothelial growth factor (VEGF) and their receptors. J Cell Sci 2001; 114(Pt 5):853-865.
24. Soker S, Takashima S, Miao HQ et al. Neuropilin-1 is expressed by endothelial and tumor cells as an isoform-specific receptor for vascular endothelial growth factor. Cell 1998; 92(6):735-745.
25. Tischer E, Mitchell R, Hartman T et al. The human gene for vascular endothelial growth factor. Multiple protein forms are encoded through alternative exon splicing. J Biol Chem 1991; 266(18):11947-11954.
26. Ng YS, Rohan R, Sunday ME et al. Differential expression of VEGF isoforms in mouse during development and in the adult. Dev Dyn 2001; 220(2):112-121.
27. Miquerol L, Gertsenstein M, Harpal K et al. Multiple developmental roles of VEGF suggested by a LacZ-tagged allele. Dev Biol 1999; 212(2):307-322.
28. Houck KA, Leung DW, Rowland AM et al. Dual regulation of vascular endothelial growth factor bioavailability by genetic and proteolytic mechanisms. J Biol Chem 1992; 267(36):26031-26037.
29. Bergers G, Brekken R, McMahon G et al. Matrix metalloproteinase-9 triggers the angiogenic switch during carcinogenesis. Nat Cell Biol 2000; 2(10):737-744.
30. Engsig MT, Chen QJ, Vu TH et al. Matrix metalloproteinase 9 and vascular endothelial growth factor are essential for osteoclast recruitment into developing long bones. J Cell Biol 2000; 151(4):879-889.
31. Iruela-Arispe ML, Carpizo D, Luque A. ADAMTS1: A matrix metalloprotease with angioinhibitory properties. Ann NY Acad Sci 2003; 995:183-190.
32. Lee S, Jilani SM, Nikolova GV et al. Processing of VEGF-A by matrix metalloproteinases regulates bioavailability and vascular patterning in tumors. J Cell Biol 2005; 169(4):681-691.
33. Whitaker GB, Limberg BJ, Rosenbaum JS. Vascular endothelial growth factor receptor-2 and neuropilin-1 form a receptor complex that is responsible for the differential signaling potency of VEGF(165) and VEGF(121). J Biol Chem 2001; 276(27):25520-25531.
34. Bachelder RE, Crago A, Chung J et al. Vascular endothelial growth factor is an autocrine survival factor for neuropilin-expressing breast carcinoma cells. Cancer Res 2001; 61(15):5736-5740.
35. Wang L, Zeng H, Wang P et al. Neuropilin-1-mediated vascular permeability factor/vascular endothelial growth factor-dependent endothelial cell migration. J Biol Chem 2003; 278(49):48848-48860.
36. Kawasaki T, Kitsukawa T, Bekku Y et al. A requirement for neuropilin-1 in embryonic vessel formation. Development 1999; 126(21):4895-4902.
37. Gerhardt H, Ruhrberg C, Abramsson A et al. Neuropilin-1 is required for endothelial tip cell guidance in the developing central nervous system. Dev Dyn 2004; 231(3):503-509.
38. Gluzman-Poltorak Z, Cohen T, Herzog Y et al. Neuropilin-2 is a receptor for the vascular endothelial growth factor (VEGF) forms VEGF-145 and VEGF-165 [corrected]. J Biol Chem 2000; 275(24):18040-18045.
39. Carmeliet P, Ng YS, Nuyens D et al. Impaired myocardial angiogenesis and ischemic cardiomyopathy in mice lacking the vascular endothelial growth factor isoforms VEGF164 and VEGF188. Nat Med 1999; 5(5):495-502.
40. Stalmans I, Ng YS, Rohan R et al. Arteriolar and venular patterning in retinas of mice selectively expressing VEGF isoforms. J Clin Invest 2002; 109(3):327-336.
41. Cheng SY, Nagane M, Huang HS et al. Intracerebral tumor-associated hemorrhage caused by overexpression of the vascular endothelial growth factor isoforms VEGF121 and VEGF165 but not VEGF189. Proc Natl Acad Sci USA 1997; 94(22):12081-12087.
42. Grunstein J, Masbad JJ, Hickey R et al. Isoforms of vascular endothelial growth factor act in a coordinate fashion To recruit and expand tumor vasculature. Mol Cell Biol 2000; 20(19):7282-7291.
43. Ruhrberg C, Gerhardt H, Golding M et al. Spatially restricted patterning cues provided by heparin-binding VEGF-A control blood vessel branching morphogenesis. Genes Dev 2002; 16(20):2684-2698.

44. Ng YS, D'Amore PA. Therapeutic angiogenesis for cardiovascular disease. Curr Control Trials Cardiovasc Med 2001; 2(6):278-285.
45. Brinkman BM. Splice variants as cancer biomarkers. Clin Biochem 2004; 37(7):584-594.
46. Ishida S, Usui T, Yamashiro K et al. VEGF164-mediated inflammation is required for pathological, but not physiological, ischemia-induced retinal neovascularization. J Exp Med 2003; 198(3):483-489.
47. Jain RK. Molecular regulation of vessel maturation. Nat Med 2003; 9(6):685-693.
48. Nakamura M, Abe Y, Tokunaga T. Pathological significance of vascular endothelial growth factor A isoform expression in human cancer. Pathol Int 2002; 52(5-6):331-339.
49. Lange T, Guttmann-Raviv N, Baruch L et al. VEGF162, a new heparin-binding vascular endothelial growth factor splice form that is expressed in transformed human cells. J Biol Chem 2003; 278(19):17164-17169.
50. Woolard J, Wang WY, Bevan HS et al. VEGF165b, an inhibitory vascular endothelial growth factor splice variant: Mechanism of action, in vivo effect on angiogenesis and endogenous protein expression. Cancer Res 2004; 64(21):7822-7835.

VEGF Receptor Signalling in Vertebrate Development

Joaquim Miguel Vieira, Christiana Ruhrberg and Quenten Schwarz*

Abstract

The secreted glycoprotein vascular endothelial growth factor A (VEGF or VEGFA) affects many different cell types and modifies a wide spectrum of cellular behaviours in tissue culture models, including proliferation, migration, differentiation and survival. The versatility of VEGF signalling is reflected in the complex composition of its cell surface receptors and their ability to activate a variety of different downstream signalling molecules. A major challenge for VEGF research is to determine which of the specific signalling pathways identified in vitro control development and homeostasis of tissues containing VEGF-responsive cell types in vivo.

Key Messages

- VEGF is expressed in different isoforms
- VEGF isoforms bind different subsets of cell surface receptors
- VEGF receptors activate a plethora of downstream signalling pathways
- VEGF receptors mediate different cellular effects

Introduction

Vascular Endothelial Growth Factor A (VEGF or VEGFA) is a critical organiser of vascular development due to its ability to regulate proliferation, migration, specialisation and survival of endothelial cells (reviewed in ref. 1). VEGF also affects many other cell types in tissue culture models. For example, it is mitogenic for lymphocytes, retinal pigment epithelium and Schwann cells.[2-4] It also stimulates the migration of haematopoietic precursors, monocytes/macrophages, neurons and vascular smooth muscle cells,[5-11] and it promotes the survival of developing and mature neurons[12] as well as chondrocytes.[13,14]

Differential splicing of the eight exons comprising the VEGF gene (*Vegfa*) gives rise to three main isoforms, termed VEGF121, VEGF165 and VEGF189 in humans and VEGF120, VEGF164 and VEGF188 in mice (see Chapter 1 by Y.-S. Ng). All VEGF isoforms bind to two type III receptor tyrosine kinases, FLT1 (*fms*-related tyrosine kinase 1, also denominated VEGFR1) and KDR (kinase insert domain containing receptor, also known as FLK1 or VEGFR2) (Fig. 1A). In contrast, heparan sulphate proteoglycans (HSPGs) and the nontyrosine kinase receptors neuropilin 1 (NRP1) and neuropilin 2 (NRP2) preferentially bind the VEGF isoforms containing the heparin-binding domains, encoded by exons 6 and 7 (Fig. 1B). In addition to the versatility provided by the existence of several different VEGF isoforms and VEGF receptors, VEGF signalling attains further plasticity from the association of VEGF receptors with other transmembrane

*Corresponding Author: Quenten Schwarz—Institute of Ophthalmology, University College London, 11-43 Bath Street, London EC1V 9EL, UK. Email: q.schwarz@ucl.ac.uk

VEGF in Development, edited by Christiana Ruhrberg. ©2008 Landes Bioscience and Springer Science+Business Media.

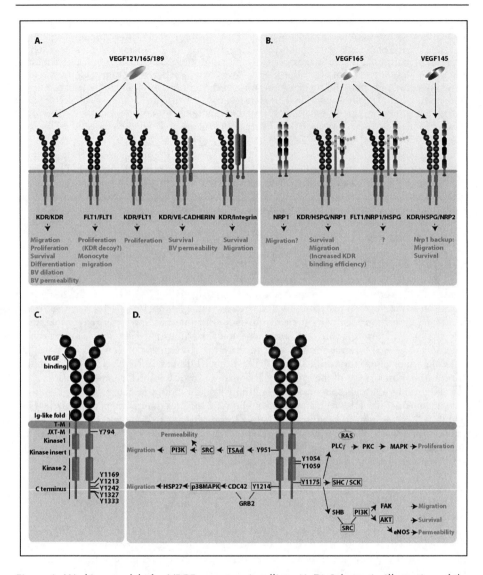

Figure 1. Working models for VEGF receptor signalling. (A-D) Schematic illustration of the different human VEGF receptors and their predicted physiological roles in endothelial cells, blood vessels and macrophages. (A) VEGF tyrosine kinase receptors: All VEGF isoforms (VEGF121, VEGF165, VEGF189) bind to homo- or heterodimers of KDR and FLT1. KDR can form higher order complexes with VE-cadherin or integrins. (B) Isoform-specific VEGF receptors: VEGF165, but not VEGF121, binds receptor complexes containing NRP1 and HSPGs, or higher order complexes containing additionally FLT1 or KDR. VEGF165 and VEGF145 bind NRP2. The neuropilin CUB domains (a1 and a2) are shown in blue, the coagulation factor V/VIII homology domains (b1 and b2) are highlighted red, and the MAM domain is coloured green. (C) FLT1 domain structure: The extracellular region consists of 7 Ig-like folds (shown as spheres); they bind ligands and mediate receptor dimerisation; the cytoplasmic domain contains two kinase domains (light brown cylinders) interrupted by a kinase insert domain; a juxtamembrane domain is thought to inhibit autophosphorylation. Legend continued on following page.

Figure 1, continued from previous page. FLT1 contains at least 7 known tyrosine phosphorylation sites (indicated by numbers that correspond to the position in the linear protein sequence); presently, in vivo data are lacking that demonstrate which of these tyrosine residues are essential for VEGF signalling in macrophages or other cell types. (D) KDR domain structure: The KDR structure is similar to that of FLT1, but KDR lacks a juxtamembrane inhibitory domain. The 7 known tyrosine phosphorylation sites are numbered according to their position in the linear protein sequence. The phosphorylated tyrosine residues are thought to interact with a collection of different proteins; experimentally confirmed interactions are represented by solid arrows, putative interactions with a dashed line. Interacting proteins that have been disrupted by gene targeting in the mouse are boxed. Abbreviation: BV, blood vessel.

proteins to form higher order signalling complexes (Fig. 1A). For example, KDR and FLT1 interact with integrins and vascular endothelial cadherin (VE-cadherin). In this chapter, we critically review current knowledge of the different VEGF signalling pathways and their interplay during development to extend a more general recent review on VEGF receptors.[15]

Tyrosine Kinase Receptors for VEGF: FLT1 and KDR

Structure of FLT1 and KDR

FLT1 and KDR are transmembrane glycoproteins of 180 and 200 kDa, respectively. They are closely related to other type III receptor tyrosine kinases, including FMS, KIT and PDGFR, and contain an extracellular domain composed of seven immunoglobulin (Ig)-like folds, a single transmembrane domain, a regulatory juxtamembrane domain and an intracellular tyrosine kinase domain (Fig. 1). The intracellular tyrosine kinase domain is interrupted by a kinase insert domain and contains several tyrosine residues that mediate the recruitment of downstream signalling molecules upon phosphorylation (Figs. 1C,D). Both KDR and FLT1 bind VEGF with high affinity. Mutation analysis of the extracellular domains of FLT1 and KDR revealed that the second and third Ig-like folds contain the high-affinity ligand-binding domain for VEGF, while the first and fourth Ig-like folds regulate ligand-binding and receptor dimerisation, respectively (Figs. 1C,D).[16-18] In addition to binding VEGF, FLT1 also acts as a receptor for VEGFB and PGF (previously known as PlGF), whilst KDR also binds the VEGF homologs VEGFC and VEGFD and the viral VEGFE.[19] Binding of VEGF by tyrosine kinase receptors promotes their homophilic or heterophilic interaction to activate the kinase domain.[20,21]

Expression Pattern of FLT1 and KDR

KDR and FLT1 are expressed in endothelial cells in most, if not all tissues in mouse and human embryos. The expression level of FLT1 in vascular endothelium varies with gestational age. Between embryonic days 8.5 and 14 (E8.5 - E14) in the mouse, the *Flt1* gene is expressed at high levels in endothelial cells, but expression decreases thereafter.[22] In newborn mice, *Flt1* expression increases again, and it continues to be expressed in adults,[22] consistent with the idea that it plays a role in the homeostasis of blood vessels. *Flt1* gene expression is regulated by hypoxia, and a binding site for hypoxia-inducible factor (HIF1A) has been identified in the *Flt1* promoter.[23] Thus, Flt1 is upregulated in vascular smooth muscle cells experiencing hypoxic stress, perhaps to control vascular remodelling or tone.[24] However, further studies are required to fully understand the physiological significance of the transcriptional regulation of FLT1 by hypoxia, and how it may complement the regulation of VEGF by hypoxia (see Chapter 3 by M. Fruttiger). In contrast to *Vegfa* and *Flt1*, *Kdr* has no HIF1A binding sites in its promoter region and is therefore not regulated by hypoxia.[23] *Kdr* is already expressed in mesodermal progenitors of vascular endothelial cells in the yolk sac at E7 in the mouse, and its expression is often used as a marker for these progenitor cells.[25-27] *Kdr* expression remains high on vascular endothelial cells during development, but it declines towards the end of gestation.[28] Nonendothelial expression of KDR has been reported in neurons, osteoblasts, pancreatic ducts cells, retinal progenitor cells, platelets and megakaryocytes (for example in refs. 29-32). Due to its expression by adult neurons after brain injury, it has been suggested that KDR has a physiological, possibly neuroprotective function (see Chapter 8 by J. Rosenstein, J. Krum and C. Ruhrberg). Like KDR, FLT1 is expressed in endothelial progenitor cells and osteoblasts, but additionally in haematopoietic stem cells, macrophages, osteoclasts, dendritic cells, pericytes, smooth muscle cells and placental trophoblasts.[33-38]

Functional Requirements for FLT1

An essential role for FLT1 in development is highlighted by the fact that FLT1-deficient mice die in utero between E8 and E9, most likely due to a failure of endothelial cells to assemble into a vascular circuit. The primary defect underlying this phenotype appears to be an altered cell fate determination among mesenchymal cells, which increases haemangioblast numbers.[39] The defect has been attributed to VEGF hyperactivity subsequent to the loss of endo-thelial FLT.[40] Two different hypotheses have been put forward to explain the negative role of FLT1 in developmental angiogenesis. The most widely accepted hypothesis suggests that FLT1 functions as a cell surface-bound "decoy receptor" to sequester excess extracellular VEGF. In support of this idea, the FLT1 kinase domain is not normally active in endothelial cells, even though FLT1 has a ten-fold higher affinity for VEGF compared to KDR; in fact, FLT1 activation in endothelial cells has only be achieved by overexpression of recombinant protein.[41-43] Moreover, mice expressing a mutant form of FLT1 with an inactive tyrosine kinase domain (*Flt1 TK-/-*) have no discernable defects in blood vessel formation, branching or remodelling, even though these mice show deficiencies in VEGF-induced macrophage migration.[44] Finally, membrane tethering of FLT1 is essential for vascular development: 50% of mice expressing solely a soluble form of FLT1, which lacks the transmembrane and tyrosine kinase domains, died between E8.5 and E9.0 with a disorganized vascular network, similar to the full knockout.[45] However, whilst 50% of mice expressing only a soluble form of FLT1 die, the other 50% of mice making only soluble FLT1 survive. A soluble form of FLT1 is produced endogenously by alternative splicing (sFLT1), raising the possibility that the soluble isoform normally cooperates with the membrane-tethered isoform to control vascular development. For example, it is conceivable that membrane bound FLT1 functions as a decoy receptor to limit VEGF availability to KDR, whilst sFLT1 sequesters soluble VEGF in the endothelial environment to sharpen VEGF gradients[46] (see Chapter 6 by H. Gerhardt).

Even though the FLT1 tyrosine kinase domain is dispensable for vascular development, FLT1 tyrosine kinase signalling significantly promotes pathological angiogenesis.[47,48] Several different explanations have been put forward to explain this difference in developmental and pathological angiogenesis. Firstly, FLT1 upregulation might activate PGF- and VEGF-responsive monocytes, which then release pro-angiogenic factors; in agreement with this idea, FLT1 tyrosine kinase signalling mediates chemotactic macrophage migration in response to PGF and VEGF,[34,35,44,49] and PGF promotes macrophage survival during tumour angiogenesis.[50] Alternatively, PGF may occupy FLT1 binding sites on endothelial cells, allowing VEGF to bind to KDR rather than FLT1; consistent with this suggestion, PGF potentiates mitogenic VEGF activity in endothelial cells in vitro, and it promotes VEGF-induced vascular permeability in vivo.[51] It is also possible that PGF binding to FLT1 promotes the transphosphorylation of KDR by FLT1 in FLT1/KDR heterodimers to increase VEGF/KDR signalling.[48] Lastly, PGF activation of FLT1 may stimulate vessel formation and maturation indirectly by acting on nonendothelial cell types, for example smooth muscle cells[24,52] or bone-marrow derived cells that are recruited to sites of neovascularisation.[25,53,54] It is presently debated whether pro-angiogenic bone-marrow derived cells support tumour angiogenesis by differentiating into endothelial cells[55] or by providing perivascular support cells.[54] The recruited perivascular cells have monocyte/macrophage characteristics, such as expression of the integrin CD11b and the hematopoietic lineage marker CD45;[54] this observation provides a link to the initial suggestion that PGF supports pathological angiogenesis by acting on cells in the monocyte/macrophage lineage. Importantly, the regulation of FLT1 by hypoxia (see above) might promote PGF responsiveness of both endothelial cells and macrophages during pathological angiogenesis.

FLT1-Stimulated Signalling Pathways

FLT1 contains several potential tyrosine autophosphorylation sites (Fig. 1C) (reviewed in ref. 56). Whereas a repressor element in the juxtamembrane domain of FLT1 inhibits auto-phosphorylation after VEGF binding,[57] this repression appears to be alleviated by an unknown mechanism in monocytes/macrophages. Biochemical assays suggest that the phosphorylated FLT1 can recruit several different proteins containing a SRC homology 2 (SH2)-domain; this domain was first identified in the SRC protein kinase. In endothelial cells, phosphorylated KDR preferentially binds to and activates SRC, whereas phosphorylated FLT1 preferentially binds two other protein kinases that are closely related to SRC, namely FYN and YES.[58] Mice lacking any one of the SRC family kinases do not suffer developmental defects, but the combined loss of SRC, FYN and YES results in embryonic lethality at E9.5.[59] Lethality may be due to vascular insufficiency downstream of KDR rather than FLT1 signalling in endothelial cells (see below). The physiological role of the different SRC family kinases in VEGF/ PGF mediated macrophage migration has not yet been examined, and the identity of the FLT1 and KDR phosphotyrosines involved in SRC kinase recruitment are also still unknown.

In addition to SRC kinase recruitment, tyrosine phosphorylation of FLT1 promotes recruitment of several other SH2 proteins, including phospholipase C gamma (PLCγ), SH2-domain containing tyrosine phosphatase 2 (SHP2), the noncatalytic region of tyrosine kinase adaptor protein 1 (NCK1), the class IA phosphatidylinositol 3-kinase (PI3K) and the cellular homolog of the viral oncogene v-crk (Fig. 1C). Phosphorylated Y1213, Y1333, Y794 and Y1169 all recruit PLCγ to activate protein kinase C (PKC). Phosphorylated Y1213 specifically binds SHP2 and NCK1. Phosphorylated Y1213 also activates PI3K, which then catalyses the production of the second messenger lipid PIP3 (Box 1). Y1333 binds CRK (the cellular homolog of v-crk) and NCK. Proteins that bind to phosphorylated Y1242 and Y1327 have so far remained elusive. Interestingly, VEGF and PGF appear to induce phosphorylation of a different subset of tyrosine residues.[48] For example, PGF, but not VEGF binding to FLT1 results in Y1309 phosphorylation and activation of the AKT cell survival pathway (see below).

Box 1. Role of class1A PI3 kinase in vascular growth. The lipid kinases of the PI3 kinase (PI3K) family produce the intracellular messenger PIP3 (phosphatidyl-inositol-3,4,5-triphosphate); one of the major functions of PIP3 is activation of the serine/threonine kinase AKT to stimulate proliferation and prevent apoptosis. The PI3Ks have been grouped into three classes, with the class I family being further subdivided into IA and IB kinases. The class 1A PI3Ks signal downstream of receptor tyrosine kinases. A role for class IA PI3Ks in endothelial cells was initially demonstrated in tissue culture models, but has more recently been studied by genetic alteration of the genes encoding its different subunits. Interpretation of the null mutant phenotypes has, however, been complicated by the fact that ablation of any one of the PI3K subunits deregulates other subunits. For example, ablation of the regulatory subunits p85a, p55a or p50 also reduces expression of the p110 catalytic subunits. Conversely, ablation of the p110a subunit results in over-expression of the p85 regulatory subunit, which has a dominant negative effect on all class IA PI3K proteins. Perhaps the most resounding evidence so far in support of an essential role for class IA PI3Ks in vascular development comes from the endothelial cell-specific knockout of PTEN (phosphatase and tensin homolog), a lipid phosphatase that reverses PI3K signalling. In this mouse model, loss of PTEN results in an overstimulation of endothelial cell proliferation and migration, causing embryonic death at E11.5.[141]

Understanding the physiological significance of the different FLT1 signalling pathways has so far proven difficult. Firstly, SHP2, PI3K, NCK and PLCγ all play roles downstream of a variety of tyrosine kinases, and the analysis of null mutants for these genes therefore cannot identify specific requirements for signalling downstream of FLT1 or KDR. Secondly, no appropriate tissue culture model with a relevant readout has been identified to evaluate the physiological importance of the different phosphorylated tyrosine residues in FLT1.[60] It would be particularly interesting to learn more about FLT1 signalling pathways in the monocyte/macrophage lineage.

Functional Requirements for KDR

Consistent with its expression in the mesodermal progenitors of blood islands in the yolk sac, *Kdr* is required for endothelial and haematopoietic cell differentiation and therefore vasculogenesis and haematopoiesis; thus, loss of KDR function results in embryonic death between E8.5 and 9.5[28] (see Chapter 4 by L. Goldie, M. Nix and K. Hirschi). As KDR is tyrosine-phosphorylated more efficiently than FLT1 upon VEGF binding in endothelial cells (see above), KDR is thought to be principally responsible for VEGF signalling to stimulate the proliferation, chemotaxis, survival, and differentiation of endothelial cells and to alter their morphology; moreover, KDR signalling is thought to stimulate vessel permeability and vessel dilation.[41,61-63] However, owing to the early lethality of *Kdr* knockout mice, the requirement for KDR in specific stages of vascular development subsequent to vasculogenesis has not yet been formally demonstrated by knockout technology.

KDR-Stimulated Signalling Pathways

KDR functions similarly to most tyrosine kinase receptors: it dimerises and is autophosphorylated on several cytoplasmic tyrosine residues upon ligand binding (Fig. 1D). Early experiments using recombinant KDR in bacteria and yeast demonstrated that several tyrosine residues are autophosphorylated upon VEGF binding to recruit SH2-domain containing proteins. The following autophosphorylated tyrosine residues were subsequently identified in human endothelial cells: in the kinase insert domain, Y951 (corresponding to Y949 in the mouse); in the tyrosine kinase domain, Y1054 and Y1059 (corresponding to Y1053 and Y1057 in the mouse); and in the C-terminal domain, Y1175 and Y1214 (corresponding to Y1173 and Y1212 in the mouse).[64] As observed in the case of FLT1, KDR phosphotyrosines are recognised by a number of different SH2-domain containing proteins. For example, SRC kinases have been implicated in signalling pathways downstream of Y951 and Y1175 (Fig. 1D), and SRC kinases modulate endothelial proliferation and migration in tissue culture models[65] and during neoangiogenesis in adults.[66] To clarify the relative contribution of the different KDR phosphotyrosines to vascular development, we will discuss the phenotypes of mice that either lack single KDR tyrosine residues or the proteins predicted to bind to them following phosphorylation.

Human Y951, Y1175 and Y1214 have all been implicated in the control of endothelial cell proliferation or migration in culture models. Y951 is selectively phosphorylated in a subset of endothelial cells during development and binds to the T cell-specific adapter molecule (TSAd), which is thought to act upstream of SRC and PI3K (Fig. 1D). Even though TSAd is critical for actin reorganization in cell culture models, it is not essential for mouse development.[67] No other protein has so far been identified that interacts functionally with Y951 in endothelial cells, and it is not known if Y951 is essential for vascular development.

Y1214 is embedded in a region of KDR that resembles the consensus binding sequence for the growth factor receptor bound protein 2 (GRB2) and has been implicated in the control of actin reorganisation and cell migration through the activation of CDC42 and the mitogen activated protein kinase (MAPK) cascade[68] (Fig. 1D). A mouse model for the tyrosine residues corresponding to human Y1214 has been created by replacing Y1212 with a phenylalanine residue; surprisingly, these mutants have no discernable defects.[69]

A mouse model for the tyrosine residue corresponding to human Y1175 has also been created by replacement of Y1173 with a phenylalanine residue. This mutation results in embryonic lethality between E8.5 and E9.5 with endothelial and haematopoietic defects, similar to those seen in complete KDR knockout mice.[70] The essential Y1175 residue, located in the KDR C-terminal domain, interacts with a number of SH2 domain-containing proteins that are expressed in endothelial cells, including PLCγ and the adaptor proteins SHCA, SHCB (also called SCK) and SHB (Fig. 1D). Activation of PLCγ leads to the activation of PKC to control endothelial cell proliferation via the MAPK pathway in cultured endothelial cells (Fig. 1D). Several different MAPK are essential for embryogenesis, with p38 and ERK5 being required for vascular development; however, it is not clear if the effects on blood vessel growth reflect a requirement in endothelial cells or occur subsequent to defective placentation[71]. SHCA KO mice suffer from embryonic lethality due to extensive vascular defects.[72] However, SHCA also interacts with other tyrosine kinase receptors that may be involved in vasculogenesis and may therefore not be a specific downstream effector of KDR. SHCB is expressed in developing blood vessels, but SHCB KO mice have no vascular defects, possibly because it acts redundantly with other SHC family members such as SHCA.[73] SHB controls endothelial cell migration through the focal adhesion kinase FAK in a pathway that involves PI3K activation (Fig. 1D). Even though SHB has not yet been knocked out in mice, it is essential for blood vessel growth in an embryonic stem cell model of angiogenesis.[74]

In addition to promoting the proliferation and migration of endothelial cells, VEGF also promotes their survival. Genetic mouse models suggest that VEGF supports endothelial cell survival in vivo by acting both in a paracrine fashion[75] and in an autocrine loop.[76] In vitro models have identified several different downstream signalling pathways that are activated by VEGF to promote endothelial survival. Paracrine survival signalling in cultured endothelial cells involves the interaction of KDR with cell adhesion molecules of the integrin family, which control cell survival in response to matrix signals in many cell types including endothelium,[77] and the interaction of KDR with VE-cadherin, a component of endothelial cell adherens junctions[78] (Fig. 1A). Mice lacking VE-cadherin die at 9.5 dpc due to vascular insufficiency, caused by defective blood vessel remodelling and maturation.[79] These defects may be due to reduced activation of anti-apoptotic protein kinases such as AKT1, a protein that promotes endothelial cell survival in vitro and in vivo.[80] AKT1 activation normally occurs downstream of VE-cadherin and VEGF/KDR in a process that requires SRC and PI3K[66,81,82] (Fig. 1D). However, AKT1 is not essential for vascular development, possibly because it signals redundantly with closely related AKT1 and AKT3 proteins. Alternatively, or additionally, VE-cadherin/KDR interaction may impact on endothelial cell survival by controlling cell surface retention of KDR.[83] It is not known which intracellular effectors play a role in autocrine VEGF survival signalling, as this pathway does not require VEGF secretion[76] and therefore is likely to bypass KDR/VE-cadherin complexes on the cell surface.

Negative Regulation of KDR Signalling

Whereas much effort has been directed at identifying the forward signalling pathways downstream of KDR, the molecular mechanisms that modulate KDR activity have received less attention. Presumably, KDR activation must be downregulated at some point to terminate signalling. The phosphatases SHP1 and SHP2 dephosphorylate the nonessential KDR Y1214 residue,[84,85] and human cellular protein tyrosine phosphatase A (HCPTPA) inhibits VEGF-signalling in tissue culture models, possibly by dephosphorylating KDR to inhibit MAPK activation.[86] Unfortunately, the physiological significance of this pathway is unknown.

VEGF Isoform-Specific Receptors: Neuropilins and HSPGs

Identification of the Neuropilins as VEGF Receptors

An isoform specific VEGF receptor that binds VEGF165, but not VEGF121, was first described in human umbilical vein-derived endothelial cells[87] and subsequently in several tumour-derived cell lines that lack the expression of other VEGF receptors.[88] This novel VEGF receptor was purified and identified as neuropilin 1 (NRP1), a 130-kDa type I transmembrane protein (Fig. 1B). NRP1 had previously been discovered as an axonal adhesion protein in the developing frog nervous system[89,90] and as a receptor for secreted guidance molecules of the class 3 semaphorin family.[91,92] Besides NRP1, the neuropilin family includes NRP2.[93] Even though NRP1 and NRP2 share only 44% homology at the amino acid level, each protein is highly conserved amongst different vertebrate species, including frog, chick, mouse and human. NRP1 and NRP2 bind a different subset of VEGF isoforms and semaphorins in vitro: whereas NRP1 preferentially binds VEGF165 and SEMA3A, NRP2 binds both VEGF165 and VEGF145 as well as SEMA3F.[91-94]

Structure of Neuropilins

NRP1 and NRP2 have an identical domain structure.[93,95] Both contain a large N-terminal extracellular domain of approximately 850 amino acids, a short membrane-spanning domain of approximately 24 amino acid residues and a small cytoplasmic domain of 40 residues. The extracellular domain contains two complement-binding (CUB) domains (termed a1 and a2), two coagulation factor V/VIII homology domains (termed b1 and b2) and a meprin (MAM) domain (Fig. 1B). The a- and b-domains are crucial for ligand binding, whilst the MAM domain promotes dimerisation and the interaction with other cell surface receptors.[96] The cytoplasmic domain is short and was originally thought to lack signalling motifs, because its deletion did not impair axonal growth cone collapse in response to SEMA3A.[97] Instead, neuropilins transduce semaphorin signals in neurons through a signalling coreceptor of the plexin family.[98,99] In analogy, it was inferred that neuropilins recruit a coreceptor such as FLT1 and KDR to transmit VEGF signals in endothelial cells. In agreement with this idea, NRP1 potentiates the signalling of coexpressed KDR in porcine aortic endothelial cells, which surprisingly lack endogenous KDR expression.[100] However, the relationship of NRP1 and KDR is different to that of NRP1 and plexins: whereas NRP1 is the compulsory ligand binding subunit in the semaphorin receptor, KDR does not require NRP1 to bind VEGF. Vice versa, recent evidence suggests that NRP1 can also signal independently of KDR in endothelial cells, suggesting that the cytoplasmic tail may have signalling activity after all (see below).

Functional Requirements for Neuropilins in Vascular Development

Neuropilins are expressed by several types of embryonic neurons, and their targeted inactivation in the mouse impairs axon guidance and neuronal migration in response to semaphorins.[101-105] In addition, loss of NRP1 disrupts neuronal migration in response to VEGF.[11]

In the vasculature, NRP1 is preferentially expressed on arterial and brain microvessel endothelium, whereas NRP2 is present on venous and lymphatic endothelium.[106,107] Consistent with a role for NRP1 in vascular growth, over-expression of NRP1 in mice deregulates angiogenesis, causing embryonic lethality at E17.5; the mutant embryos exhibit excess capillaries and blood vessels, dilation of blood vessels, severe haemorrhage and malformed hearts.[107] Mice lacking NRP1 die even earlier, at around E12.5, with impaired neural tube vascularisation, agenesis or transposition of the aortic arches, persistent truncus arteriosus and insufficient development of the yolk sac vasculature.[108] The physiological role of NRP1 during vascular development has also been addressed in zebrafish models. In this organism, knockdown of NRP1 impairs angiogenic sprouting from the major axial vessels and therefore formation of the intersomitic vessels.[109] Others have shown that the knockdown of NRP1 in zebrafish disrupts even earlier stages of vascular development, including the formation of the dorsal longitudinal anastomosing vessels and the subintestinal vein.[110]

Consistent with its expression pattern, mice lacking NRP2 are deficient in the formation of small lymphatics and capillaries, but they show no other obvious cardiovascular abnormalities.[111] Noteworthy, loss of both NRP1 and NRP2 in mice impairs vascular development more severely than loss of NRP1 alone, with death at E8.0 due to impaired yolk sac vascularisation;[112] these data suggest that both proteins can partially compensate for each other during the formation of arterio-venous circuits. However, the reason why NRP2 is able to compensate for NRP1 during vascular development is presently unclear. The observation that both proteins are expressed in a reciprocal pattern during the segregation of arterio-venous circuit in the chick[106] raises the possibility that venous NRP2 function becomes essential only when arterial NRP1 expression is lost. Alternatively, NRP2 may be upregulated in NRP1-deficient vascular endothelial cells to compensate for NRP1. Consistent with this hypothesis, NRP2 is able to enhance KDR signalling in porcine aortic endothelial cells,[113] which lack NRP1.[100]

The requirement for NRP1 in vascular growth is generally considered to reflect its essential role in promoting VEGF165 signalling in endothelial cells. In agreement with this idea, mice lacking NRP1 specifically in vascular endothelium show impaired microvessel growth in the brain.[114] However, there are some striking differences in the vascular defects caused by loss of NRP1 or loss of its VEGF ligands in the developing trunk and central nervous system, with loss of NRP1 causing a more severe vascular deficiency particularly in the brain.[108,115,116] These observations suggest that loss of VEGF isoform signalling through NRP1 is not entirely responsible for the vascular deficiency of NRP1 null mutants, and that NRP1 ligands other than VEGF165 may contribute to vessel patterning. The finding that SEMA3A inhibits endothelial cell migration in vitro by competing with VEGF165 for binding to NRP1/KDR complexes made it a candidate modulator of neuropilin-mediated vessel patterning in vivo.[117] Yet, class 3 semaphorin-signalling through neuropilins is not required for embryonic vascular development.[114,118] Therefore, the nature of the hypothetical NRP1 ligand that cooperates with VEGF165 during vascular patterning remains elusive.

VEGF165/NRP1 Signalling

In analogy to the compulsory recruitment of plexins to transmit semaphorin signals, NRP1 was initially proposed to recruit a coreceptor such as FLT1 and KDR to transmit VEGF165 signals.[100,119,120] In support of this idea, NRP1 does not promote the VEGF165-induced chemotaxis of KDR-negative cultured porcine aortic endothelial cells, but when coexpressed with KDR, it enhances chemotaxis more than KDR alone.[100] Two alternative hypotheses have been proposed to explain the beneficial effect of NRP1 on KDR signalling: Complexes containing both KDR and NRP1 may bind VEGF165 with higher affinity than KDR or NRP1 alone,[121] or NRP1 may promote KDR clustering to promote VEGF165 signalling.[120]

However, other tissue culture models suggest that NRP1 may also function in endothelial cells independently of its ability to enhance VEGF/KDR signalling. Firstly, when the extracellular domain of epidermal growth factor (EGF) receptor was fused to a NRP1 fragment comprised of its membrane-spanning and cytoplasmic domain, the chimeric receptor promoted endothelial cell migration in response to EGF.[122] Secondly, the last three amino acid residues of NRP1 (SEA-COOH) bind to the neuropilin-interacting protein NIP (also known as GIPC or synectin),[123] and this interaction contributes to vascular development in zebrafish and mice.[113,124] One zebrafish study demonstrated that disruption of trunk vessel development by NRP1 knock down could be rescued by delivery of full length human NRP1, but not by human NRP1 lacking the NIP-binding SEA motif.[113] Moreover, ectopic expression of NRP1 lacking the SEA motif or knockdown of NIP disrupted vessel growth in this study.[113] Another zebrafish study found that knockdown of NIP affected vascular development at an even earlier stage by impairing dorsal aorta formation.[124] In mice, loss of NIP leads to less severe cardiovascular defects than loss of NRP1.[124] NIP null mice are born at the expected Mendelian frequency; moreover, the brain and spinal chord are vascularised normally, even though these tissues are severely affected in NRP1 null mutants (J. M. V., C. R. and M. Simmons, unpublished observations). However, NIP-deficient mice show a specific defect in arterial development and adult arteriogenesis, with reduced arterial density and branching in the retina, heart and kidney.[124] These observations agree with those of other mouse studies, in which loss of VEGF165 signalling through NRP1 affected arterial patterning in the limb skin[125] and in the retina.[126]

In summary, NRP1 is likely to play a dual role in vascular growth by enhancing VEGF164 signalling through KDR and by promoting VEGF164 signalling through its own intracellular domain.

Heparan Sulphate Proteoglycans

Heparan sulphate proteoglycans (HSPGs) are abundant and highly conserved components of the cell surface and extracellular matrix. They play an important role in the formation and modulation of gradients of heparin-binding growth factors, morphogens and chemokines.[127] Several reports have implicated HSPGs as modulators of VEGF signalling. Firstly, VEGF164 and VEGF188 bind heparin in vitro with different degrees of affinity, depending on the presence/absence of the so-called heparin-binding domains; heparin-binding ability in vitro is thought to indicate HSPG binding in vivo.[128] In support of this idea, loss of the heparin-binding VEGF isoforms affects VEGF distribution in the extracellular matrix during angiogenic sprouting in the brain and retina.[115,129] Heparin also promotes VEGF165 binding to its receptors KDR[130,131] and NRP1.[88,100] Moreover, when heparan sulphate is enzymatically removed from endothelial cells, KDR phosphorylation is inhibited.[132] The beneficial effect of heparin or heparan sulphate on VEGF signalling may additionally stem from a direct interaction with the VEGF receptors. Consistent with this suggestion, HSPG expression by perivascular smooth muscle cells transactivates endothelial KDR in an embryonic stem cell model of angiogenesis, and possibly facilitates the cross talk between both cell types during blood vessel formation in vivo.[133] Finally, NRP1 may itself become a proteoglycan by post-transcriptional modification with glycosaminglycan side chains of the heparan sulphate or the chondroitin sulphate type, and this modification may enhance VEGF binding.[134]

Conclusions and Future Perspectives

Initially, VEGF signalling pathways were characterised in tissue culture models of endothelium. More recently, the physiological relevance of the different pathways has been addressed with mouse mutants that carry point mutations in single KDR tyrosine phosphorylation sites or harbour null mutations in proteins that interact with these tyrosines. A more complete understanding of KDR signalling will, however, depend on the creation of further mouse mutants lacking other KDR tyrosine residues implicated in intracellular signalling, as well as the design of a novel strategy to study FLT1 tyrosine kinase signalling in vivo.

Despite the progress made in identifying intracellular adaptor molecules for KDR and FLT1, we still know very little about the intracellular trafficking of VEGF and its receptor complexes. For example, in some endothelial culture models the phosphorylation of tyrosine residues Y1054 and Y1059 controls internalisation of the VEGF/KDR complex into clathrin-coated vesicles and endosomes prior to degradation,[135] and KDR may also signal from endosomes to promote endothelial cell proliferation.[83] In other endothelial tissue culture models, VEGF stimulates nuclear translocation of KDR.[136-139] In addition, circumstantial evidence is emerging that autocrine VEGF signalling may be based on intracrine signalling; for example, autocrine VEGF survival signalling in endothelial cells does not require VEGF secretion.[76] Further effort should therefore be directed at establishing the physiological significance of intracellular interactions between VEGF and its receptors during development or disease.

Owing to the absolute requirement for VEGF during embryogenesis, many previous studies focussed on elucidating the physiological requirement for VEGF signalling pathways in early vascular development. These studies also benefited from the fact that developmental angiogenesis produces a stereotypic pattern of hierarchical blood vessel networks in a well-defined tissue context. In contrast, adult angiogenesis occurs against a backdrop of environmental fluctuations and is influenced by the dynamic interaction of growing vessels with the immune system. Nevertheless, research into developmental VEGF signalling pathways has impacted on our understanding of neoangiogenesis in the adult, owing to the reactivation of VEGF signalling pathways in physiological processes such as wound healing and pathological conditions such as cancer, diabetic retinopathy and ischemic heart disease. Thus, the potential of novel anti-angiogenic therapies can be evaluated in the perinatal rodent eye before being tested in a disease model, because the rodent retina is vascularised only after birth and the eye is easily accessible for drug delivery (e.g., ref. 140).

Finally, it will be necessary to extend the study of VEGF signalling pathways to include other VEGF-responsive cell types, most notably circulating progenitors cells (see Chapter 4 by L. C. Goldie, M. K. Nix and K. K. Hirschi), bone cell types (see Chapter 7 by C. Maes and G. Carmeliet) and neuronal progenitors (see Chapter 8 by J. M. Rosenstein, J. M. Krum and C. Ruhrberg). It will be particularly interesting to elucidate if different VEGF-responsive cell types that grow in close spatiotemporal proximity activate distinct VEGF signalling pathways to coordinate their behaviour. For example, VEGF signalling is likely to play a dual role in blood vessels and bone cell types during bone development (see Chapter 7 by C. Maes and G. Carmeliet), and it supports both blood vessel growth and neuronal growth in the angiogenic niche of neurogenesis (see Chapter 8 by J. Rosenstein, J. Krum and C. Ruhrberg). The identification of cell-type specific components in the VEGF signalling pathway might then provide the basis for the creation of selective tools to balance vascular effects of VEGF such as permeability against effects on nonendothelial cell types in novel pro- and anti-angiogenic therapies.

Acknowledgements

We thank Dr Patric Turowski for critical reading of the manuscript. J. M. V. holds a PhD studentship from the Fundação para a Ciência e Tecnologia (SFRH/BD/17812/2004). Q. S. is supported by an MRC project grant (G0600993).

References

1. Ruhrberg C. Growing and shaping the vascular tree: Multiple roles for VEGF. Bioessays 2003; 25(11):1052-1060.
2. Praloran V. Structure, biosynthesis and biological roles of monocyte-macrophage colony stimulating factor (CSF-1 or M-CSF). Nouv Rev Fr Hematol 1991; 33(4):323-333.
3. Guerrin M, Moukadiri H, Chollet P et al. Vasculotropin/vascular endothelial growth factor is an autocrine growth factor for human retinal pigment epithelial cells cultured in vitro. J Cell Physiol 1995; 164(2):385-394.
4. Sondell M, Lundborg G, Kanje M. Vascular endothelial growth factor has neurotrophic activity and stimulates axonal outgrowth, enhancing cell survival and Schwann cell proliferation in the peripheral nervous system. J Neurosci 1999; 19(14):5731-5740.
5. Grosskreutz CL, Anand-Apte B, Duplaa C et al. Vascular endothelial growth factor-induced migration of vascular smooth muscle cells in vitro. Microvasc Res 1999; 58(2):128-136.
6. Wang H, Keiser JA. Vascular endothelial growth factor upregulates the expression of matrix metalloproteinases in vascular smooth muscle cells: Role of flt-1. Circ Res 1998; 83(8):832-840.
7. Clauss M, Gerlach M, Gerlach H et al. Vascular permeability factor: A tumor-derived polypeptide that induces endothelial cell and monocyte procoagulant activity, and promotes monocyte migration. J Exp Med 1990; 172(6):1535-1545.
8. Senger DR, Ledbetter SR, Claffey KP et al. Stimulation of endothelial cell migration by vascular permeability factor/vascular endothelial growth factor through cooperative mechanisms involving the alphavbeta3 integrin, osteopontin, and thrombin. Am J Pathol 1996; 149(1):293-305.
9. Cleaver O, Krieg PA. VEGF mediates angioblast migration during development of the dorsal aorta in Xenopus. Development 1998; 125(19):3905-3914.
10. Shalaby F, Ho J, Stanford WL et al. A requirement for Flk1 in primitive and definitive hematopoiesis and vasculogenesis. Cell 1997; 89(6):981-990.
11. Schwarz Q, Gu C, Fujisawa H et al. Vascular endothelial growth factor controls neuronal migration and cooperates with Sema3A to pattern distinct compartments of the facial nerve. Genes Dev 2004; 18(22):2822-2834.
12. Silverman WF, Krum JM, Mani N et al. Vascular, glial and neuronal effects of vascular endothelial growth factor in mesencephalic explant cultures. Neuroscience 1999; 90(4):1529-1541.
13. Zelzer E, Mamluk R, Ferrara N et al. VEGFA is necessary for chondrocyte survival during bone development. Development 2004; 131(9):2161-2171.
14. Maes C, Stockmans I, Moermans K et al. Soluble VEGF isoforms are essential for establishing epiphyseal vascularization and regulating chondrocyte development and survival. J Clin Invest 2004; 113(2):188-199.

15. Olsson AK, Dimberg A, Kreuger J et al. VEGF receptor signalling—In control of vascular function. Nat Rev Mol Cell Biol 2006; 7(5):359-371.
16. Davis-Smyth T, Presta LG, Ferrara N. Mapping the charged residues in the second immunoglobulin-like domain of the vascular endothelial growth factor/placenta growth factor receptor Flt-1 required for binding and structural stability. J Biol Chem 1998; 273(6):3216-3222.
17. Fuh G, Li B, Crowley C et al. Requirements for binding and signaling of the kinase domain receptor for vascular endothelial growth factor. J Biol Chem 1998; 273(18):11197-11204.
18. Shinkai A, Ito M, Anazawa H et al. Mapping of the sites involved in ligand association and dissociation at the extracellular domain of the kinase insert domain-containing receptor for vascular endothelial growth factor. J Biol Chem 1998; 273(47):31283-31288.
19. Cebe-Suarez S, Zehnder-Fjallman A, Ballmer-Hofer K. The role of VEGF receptors in angiogenesis; complex partnerships. Cell Mol Life Sci 2006; 63(5):601-615.
20. Kendall RL, Wang G, DiSalvo J et al. Specificity of vascular endothelial cell growth factor receptor ligand binding domains. Biochem Biophys Res Commun 1994; 201(1):326-330.
21. Huang K, Andersson C, Roomans GM et al. Signaling properties of VEGF receptor-1 and -2 homo- and heterodimers. Int J Biochem Cell Biol 2001; 33(4):315-324.
22. Peters KG, De Vries C, Williams LT. Vascular endothelial growth factor receptor expression during embryogenesis and tissue repair suggests a role in endothelial differentiation and blood vessel growth. Proc Natl Acad Sci USA 1993; 90(19):8915-8919.
23. Gerber HP, Condorelli F, Park J et al. Differential transcriptional regulation of the two vascular endothelial growth factor receptor genes. Flt-1, but not Flk-1/KDR, is up-regulated by hypoxia. J Biol Chem 1997; 272(38):23659-23667.
24. Bellik L, Vinci MC, Filippi S et al. Intracellular pathways triggered by the selective FLT-1-agonist placental growth factor in vascular smooth muscle cells exposed to hypoxia. Br J Pharmacol 2005; 146(4):568-575.
25. Hattori K, Heissig B, Wu Y et al. Placental growth factor reconstitutes hematopoiesis by recruiting VEGFR1(+) stem cells from bone-marrow microenvironment. Nat Med 2002; 8(8):841-849.
26. Bellamy WT. Vascular endothelial growth factor as a target opportunity in hematological malignancies. Curr Opin Oncol 2002; 14(6):649-656.
27. Ishida A, Murray J, Saito Y et al. Expression of vascular endothelial growth factor receptors in smooth muscle cells. J Cell Physiol 2001; 188(3):359-368.
28. Shalaby F, Rossant J, Yamaguchi TP et al. Failure of blood-island formation and vasculogenesis in Flk-1-deficient mice. Nature 1995; 376(6535):62-66.
29. Ogunshola OO, Antic A, Donoghue MJ et al. Paracrine and autocrine functions of neuronal vascular endothelial growth factor (VEGF) in the central nervous system. J Biol Chem 2002; 277(13):11410-11415.
30. Yang X, Cepko CL. Flk-1, a receptor for vascular endothelial growth factor (VEGF), is expressed by retinal progenitor cells. J Neurosci 1996; 16(19):6089-6099.
31. Casella I, Feccia T, Chelucci C et al. Autocrine-paracrine VEGF loops potentiate the maturation of megakaryocytic precursors through Flt1 receptor. Blood 2003; 101(4):1316-1323.
32. Selheim F, Holmsen H, Vassbotn FS. Identification of functional VEGF receptors on human platelets. FEBS Lett 2002; 512(1-3):107-110.
33. Sawano A, Iwai S, Sakurai Y et al. Flt-1, vascular endothelial growth factor receptor 1, is a novel cell surface marker for the lineage of monocyte-macrophages in humans. Blood 2001; 97(3):785-791.
34. Clauss M, Weich H, Breier G et al. The vascular endothelial growth factor receptor Flt-1 mediates biological activities. Implications for a functional role of placenta growth factor in monocyte activation and chemotaxis. J Biol Chem 1996; 271(30):17629-17634.
35. Barleon B, Sozzani S, Zhou D et al. Migration of human monocytes in response to vascular endothelial growth factor (VEGF) is mediated via the VEGF receptor flt-1. Blood 1996; 87(8):3336-3343.
36. Dikov MM, Ohm JE, Ray N et al. Differential roles of vascular endothelial growth factor receptors 1 and 2 in dendritic cell differentiation. J Immunol 2005; 174(1):215-222.
37. Nomura M, Yamagishi S, Harada S et al. Possible participation of autocrine and paracrine vascular endothelial growth factors in hypoxia-induced proliferation of endothelial cells and pericytes. J Biol Chem 1995; 270(47):28316-28324.
38. Kaipainen A, Korhonen J, Pajusola K et al. The related FLT4, FLT1, and KDR receptor tyrosine kinases show distinct expression patterns in human fetal endothelial cells. J Exp Med 1993; 178(6):2077-2088.
39. Fong GH, Rossant J, Gertsenstein M et al. Role of the Flt-1 receptor tyrosine kinase in regulating the assembly of vascular endothelium. Nature 1995; 376(6535):66-70.
40. Fong GH, Zhang L, Bryce DM et al. Increased hemangioblast commitment, not vascular disorganization, is the primary defect in flt-1 knock-out mice. Development 1999; 126(13):3015-3025.

41. Waltenberger J, Claesson-Welsh L, Siegbahn A et al. Different signal transduction properties of KDR and Flt1, two receptors for vascular endothelial growth factor. J Biol Chem 1994; 269(43):26988-26995.
42. Olofsson B, Korpelainen E, Pepper MS et al. Vascular endothelial growth factor B (VEGF-B) binds to VEGF receptor-1 and regulates plasminogen activator activity in endothelial cells. Proc Natl Acad Sci USA 1998; 95(20):11709-11714.
43. Gille H, Kowalski J, Li B et al. Analysis of biological effects and signaling properties of Flt-1 (VEGFR-1) and KDR (VEGFR-2): A reassessment using novel receptor-specific vascular endothelial growth factor mutants. J Biol Chem 2001; 276(5):3222-3230.
44. Hiratsuka S, Minowa O, Kuno J et al. Flt-1 lacking the tyrosine kinase domain is sufficient for normal development and angiogenesis in mice. Proc Natl Acad Sci USA 1998; 95(16):9349-9354.
45. Hiratsuka S, Nakao K, Nakamura K et al. Membrane fixation of vascular endothelial growth factor receptor 1 ligand-binding domain is important for vasculogenesis and angiogenesis in mice. Mol Cell Biol 2005; 25(1):346-354.
46. Kearney JB, Kappas NC, Ellerstrom C et al. The VEGF receptor flt-1 (VEGFR-1) is a positive modulator of vascular sprout formation and branching morphogenesis. Blood 2004; 103(12):4527-4535.
47. Hiratsuka S, Maru Y, Okada A et al. Involvement of Flt-1 tyrosine kinase (vascular endothelial growth factor receptor-1) in pathological angiogenesis. Cancer Res 2001; 61(3):1207-1213.
48. Autiero M, Waltenberger J, Communi D et al. Role of PlGF in the intra- and intermolecular cross talk between the VEGF receptors Flt1 and Flk1. Nat Med 2003; 9(7):936-943.
49. Selvaraj SK, Giri RK, Perelman N et al. Mechanism of monocyte activation and expression of proinflammatory cytochemokines by placenta growth factor. Blood 2003; 102(4):1515-1524.
50. Adini A, Kornaga T, Firoozbakht F et al. Placental growth factor is a survival factor for tumor endothelial cells and macrophages. Cancer Res 2002; 62(10):2749-2752.
51. Park JE, Chen HH, Winer J et al. Placenta growth factor. Potentiation of vascular endothelial growth factor bioactivity, in vitro and in vivo, and high affinity binding to Flt-1 but not to Flk-1/KDR. J Biol Chem 1994; 269(41):25646-25654.
52. Parenti A, Brogelli L, Filippi S et al. Effect of hypoxia and endothelial loss on vascular smooth muscle cell responsiveness to VEGF-A: Role of flt-1/VEGF-receptor-1. Cardiovasc Res 2002; 55(1):201-212.
53. Luttun A, Tjwa M, Moons L et al. Revascularization of ischemic tissues by PlGF treatment, and inhibition of tumor angiogenesis, arthritis and atherosclerosis by anti-Flt1. Nat Med 2002; 8(8):831-840.
54. Grunewald M, Avraham I, Dor Y et al. VEGF-induced adult neovascularization: Recruitment, retention, and role of accessory cells. Cell 2006; 124(1):175-189.
55. Lyden D, Hattori K, Dias S et al. Impaired recruitment of bone-marrow-derived endothelial and hematopoietic precursor cells blocks tumor angiogenesis and growth. Nat Med 2001; 7(11):1194-1201.
56. Matsumoto T, Claesson-Welsh L. VEGF receptor signal transduction. Sci STKE 2001; 2001(112):RE21.
57. Gille H, Kowalski J, Yu L et al. A repressor sequence in the juxtamembrane domain of Flt-1 (VEGFR-1) constitutively inhibits vascular endothelial growth factor-dependent phosphatidylinositol 3'-kinase activation and endothelial cell migration. EMBO J 2000; 19(15):4064-4073.
58. Chou MT, Wang J, Fujita DJ. Src kinase becomes preferentially associated with the VEGFR, KDR/Flk-1, following VEGF stimulation of vascular endothelial cells. BMC Biochem 2002; 3:32.
59. Stein PL, Vogel H, Soriano P. Combined deficiencies of Src, Fyn, and Yes tyrosine kinases in mutant mice. Genes Dev 1994; 8(17):1999-2007.
60. Ito N, Huang K, Claesson-Welsh L. Signal transduction by VEGF receptor-1 wild type and mutant proteins. Cell Signal 2001; 13(11):849-854.
61. Terman BI, Dougher-Vermazen M, Carrion ME et al. Identification of the KDR tyrosine kinase as a receptor for vascular endothelial cell growth factor. Biochem Biophys Res Commun 1992; 187(3):1579-1586.
62. Quinn TP, Peters KG, De Vries C et al. Fetal liver kinase 1 is a receptor for vascular endothelial growth factor and is selectively expressed in vascular endothelium. Proc Natl Acad Sci USA 1993; 90(16):7533-7537.
63. Bernatchez PN, Soker S, Sirois MG. Vascular endothelial growth factor effect on endothelial cell proliferation, migration, and platelet-activating factor synthesis is Flk-1-dependent. J Biol Chem 1999; 274(43):31047-31054.
64. Takahashi T, Yamaguchi S, Chida K et al. A single autophosphorylation site on KDR/Flk-1 is essential for VEGF-A-dependent activation of PLC-gamma and DNA synthesis in vascular endothelial cells. Embo J 2001; 20(11):2768-2778.

65. Werdich XQ, Penn JS. Src, Fyn and Yes play differential roles in VEGF-mediated endothelial cell events. Angiogenesis 2005; 8(4):315-326.
66. Eliceiri BP, Paul R, Schwartzberg PL et al. Selective requirement for Src kinases during VEGF-induced angiogenesis and vascular permeability. Mol Cell 1999; 4(6):915-924.
67. Rajagopal K, Sommers CL, Decker DC et al. RIBP, a novel Rlk/Txk- and itk-binding adaptor protein that regulates T cell activation. J Exp Med 1999; 190(11):1657-1668.
68. Lamalice L, Houle F, Jourdan G et al. Phosphorylation of tyrosine 1214 on VEGFR2 is required for VEGF-induced activation of Cdc42 upstream of SAPK2/p38. Oncogene 2004; 23(2):434-445.
69. Holmqvist K, Cross MJ, Rolny C et al. The adaptor protein shb binds to tyrosine 1175 in vascular endothelial growth factor (VEGF) receptor-2 and regulates VEGF-dependent cellular migration. J Biol Chem 2004; 279(21):22267-22275.
70. Sakurai Y, Ohgimoto K, Kataoka Y et al. Essential role of Flk-1 (VEGF receptor 2) tyrosine residue 1173 in vasculogenesis in mice. Proc Natl Acad Sci USA 2005; 102(4):1076-1081.
71. Aouadi M, Binetruy B, Caron L et al. Role of MAPKs in development and differentiation: Lessons from knockout mice. Biochimie 2006; 88(9):1091-1098.
72. Lai KM, Pawson T. The ShcA phosphotyrosine docking protein sensitizes cardiovascular signaling in the mouse embryo. Genes Dev 2000; 14(9):1132-1145.
73. Sakai R, Henderson JT, O'Bryan JP et al. The mammalian ShcB and ShcC phosphotyrosine docking proteins function in the maturation of sensory and sympathetic neurons. Neuron 2000; 28(3):819-833.
74. Kriz V, Agren N, Lindholm CK et al. The SHB adapter protein is required for normal maturation of mesoderm during in vitro differentiation of embryonic stem cells. J Biol Chem 2006; 281(45):34484-34491.
75. Haigh JJ, Morelli PI, Gerhardt H et al. Cortical and retinal defects caused by dosage-dependent reductions in VEGF-A paracrine signaling. Dev Biol 2003; 262(2):225-241.
76. Lee S, Chen TT, Barber CL et al. Autocrine VEGF signaling is required for vascular homeostasis. Cell 2007; 130(4):691-703.
77. Serini G, Valdembri D, Bussolino F. Integrins and angiogenesis: A sticky business. Exp Cell Res 2006; 312(5):651-658.
78. Carmeliet P, Collen D. Molecular basis of angiogenesis: Role of VEGF and VE-cadherin. Ann NY Acad Sci 2000; 902:249-262, (discussion 262-244).
79. Carmeliet P, Lampugnani MG, Moons L et al. Targeted deficiency or cytosolic truncation of the VE-cadherin gene in mice impairs VEGF-mediated endothelial survival and angiogenesis. Cell 1999; 98(2):147-157.
80. Sun JF, Phung T, Shiojima I et al. Microvascular patterning is controlled by fine-tuning the Akt signal. Proc Natl Acad Sci USA 2005; 102(1):128-133.
81. Gerber HP, McMurtrey A, Kowalski J et al. Vascular endothelial growth factor regulates endothelial cell survival through the phosphatidylinositol 3'-kinase/Akt signal transduction pathway. Requirement for Flk-1/KDR activation. J Biol Chem 1998; 273(46):30336-30343.
82. Thakker GD, Hajjar DP, Muller WA et al. The role of phosphatidylinositol 3-kinase in vascular endothelial growth factor signaling. J Biol Chem 1999; 274(15):10002-10007.
83. Lampugnani MG, Orsenigo F, Gagliani MC et al. Vascular endothelial cadherin controls VEGFR-2 internalization and signaling from intracellular compartments. J Cell Biol 2006; 174(4):593-604.
84. Guo DQ, Wu LW, Dunbar JD et al. Tumor necrosis factor employs a protein-tyrosine phosphatase to inhibit activation of KDR and vascular endothelial cell growth factor-induced endothelial cell proliferation. J Biol Chem 2000; 275(15):11216-11221.
85. Gallicchio M, Mitola S, Valdembri D et al. Inhibition of vascular endothelial growth factor receptor 2-mediated endothelial cell activation by Axl tyrosine kinase receptor. Blood 2005; 105(5):1970-1976.
86. Huang L, Sankar S, Lin C et al. HCPTPA, a protein tyrosine phosphatase that regulates vascular endothelial growth factor receptor-mediated signal transduction and biological activity. J Biol Chem 1999; 274(53):38183-38188.
87. Gitay-Goren H, Cohen T, Tessler S et al. Selective binding of VEGF121 to one of the three vascular endothelial growth factor receptors of vascular endothelial cells. J Biol Chem 1996; 271(10):5519-5523.
88. Soker S, Fidder H, Neufeld G et al. Characterization of novel vascular endothelial growth factor (VEGF) receptors on tumor cells that bind VEGF165 via its exon 7-encoded domain. J Biol Chem 1996; 271(10):5761-5767.
89. Takagi S, Tsuji T, Amagai T et al. Specific cell surface labels in the visual centers of Xenopus laevis tadpole identified using monoclonal antibodies. Dev Biol 1987; 122(1):90-100.
90. Fujisawa H, Ohtsuki T, Takagi S et al. An aberrant retinal pathway and visual centers in Xenopus tadpoles share a common cell surface molecule, A5 antigen. Dev Biol 1989; 135(2):231-240.

91. He Z, Tessier-Lavigne M. Neuropilin is a receptor for the axonal chemorepellent Semaphorin III. Cell 1997; 90(4):739-751.
92. Kolodkin AL, Levengood DV, Rowe EG et al. Neuropilin is a semaphorin III receptor. Cell 1997; 90(4):753-762.
93. Chen H, Chedotal A, He Z et al. Neuropilin-2, a novel member of the neuropilin family, is a high affinity receptor for the semaphorins Sema E and Sema IV but not Sema III. Neuron 1997; 19(3):547-559.
94. Gluzman-Poltorak Z, Cohen T, Herzog Y et al. Neuropilin-2 is a receptor for the vascular endothelial growth factor (VEGF) forms VEGF-145 and VEGF-165 [corrected]. J Biol Chem 2000; 275(24):18040-18045.
95. Takagi S, Hirata T, Agata K et al. The A5 antigen, a candidate for the neuronal recognition molecule, has homologies to complement components and coagulation factors. Neuron 1991; 7(2):295-307.
96. Lee CC, Kreusch A, McMullan D et al. Crystal structure of the human neuropilin-1 b1 domain. Structure 2003; 11(1):99-108.
97. Nakamura F, Tanaka M, Takahashi T et al. Neuropilin-1 extracellular domains mediate semaphorin D/III-induced growth cone collapse. Neuron 1998; 21(5):1093-1100.
98. Rohm B, Ottemeyer A, Lohrum M et al. Plexin/neuropilin complexes mediate repulsion by the axonal guidance signal semaphorin 3A. Mech Dev 2000; 93(1-2):95-104.
99. Tamagnone L, Artigiani S, Chen H et al. Plexins are a large family of receptors for transmembrane, secreted, and GPI-anchored semaphorins in vertebrates. Cell 1999; 99(1):71-80.
100. Soker S, Takashima S, Miao HQ et al. Neuropilin-1 is expressed by endothelial and tumor cells as an isoform-specific receptor for vascular endothelial growth factor. Cell 1998; 92(6):735-745.
101. Kitsukawa T, Shimizu M, Sanbo M et al. Neuropilin-semaphorin III/D-mediated chemorepulsive signals play a crucial role in peripheral nerve projection in mice. Neuron 1997; 19(5):995-1005.
102. Kawasaki T, Bekku Y, Suto F et al. Requirement of neuropilin 1-mediated Sema3A signals in patterning of the sympathetic nervous system. Development 2002; 129(3):671-680.
103. Marin O, Yaron A, Bagri A et al. Sorting of striatal and cortical interneurons regulated by semaphorin-neuropilin interactions. Science 2001; 293(5531):872-875.
104. Giger RJ, Cloutier JF, Sahay A et al. Neuropilin-2 is required in vivo for selective axon guidance responses to secreted semaphorins. Neuron 2000; 25(1):29-41.
105. Chen H, Bagri A, Zupicich JA et al. Neuropilin-2 regulates the development of selective cranial and sensory nerves and hippocampal mossy fiber projections. Neuron 2000; 25(1):43-56.
106. Herzog Y, Kalcheim C, Kahane N et al. Differential expression of neuropilin-1 and neuropilin-2 in arteries and veins. Mech Dev 2001; 109(1):115-119.
107. Kitsukawa T, Shimono A, Kawakami A et al. Overexpression of a membrane protein, neuropilin, in chimeric mice causes anomalies in the cardiovascular system, nervous system and limbs. Development 1995; 121(12):4309-4318.
108. Kawasaki T, Kitsukawa T, Bekku Y et al. A requirement for neuropilin-1 in embryonic vessel formation. Development 1999; 126(21):4895-4902.
109. Lee P, Goishi K, Davidson AJ et al. Neuropilin-1 is required for vascular development and is a mediator of VEGF-dependent angiogenesis in zebrafish. Proc Natl Acad Sci USA 2002; 99(16):10470-10475.
110. Wang L, Mukhopadhyay D, Xu X. C terminus of RGS-GAIP-interacting protein conveys neuropilin-1-mediated signaling during angiogenesis. FASEB J 2006; 20(9):1513-1515.
111. Yuan L, Moyon D, Pardanaud L et al. Abnormal lymphatic vessel development in neuropilin 2 mutant mice. Development 2002; 129(20):4797-4806.
112. Takashima S, Kitakaze M, Asakura M et al. Targeting of both mouse neuropilin-1 and neuropilin-2 genes severely impairs developmental yolk sac and embryonic angiogenesis. Proc Natl Acad Sci USA 2002; 99(6):3657-3662.
113. Favier B, Alam A, Barron P et al. Neuropilin-2 interacts with VEGFR-2 and VEGFR-3 and promotes human endothelial cell survival and migration. Blood 2006; 108(4):1243-1250.
114. Gu C, Rodriguez ER, Reimert DV et al. Neuropilin-1 conveys semaphorin and VEGF signaling during neural and cardiovascular development. Dev Cell 2003; 5(1):45-57.
115. Ruhrberg C, Gerhardt H, Golding M et al. Spatially restricted patterning cues provided by heparin-binding VEGF-A control blood vessel branching morphogenesis. Genes Dev 2002; 16(20):2684-2698.
116. Gerhardt H, Ruhrberg C, Abramsson A et al. Neuropilin-1 is required for endothelial tip cell guidance in the developing central nervous system. Dev Dyn 2004; 231(3):503-509.

117. Miao HQ, Soker S, Feiner L et al. Neuropilin-1 mediates collapsin-1/semaphorin III inhibition of endothelial cell motility: Functional competition of collapsin-1 and vascular endothelial growth factor-165. J Cell Biol 1999; 146(1):233-242.

118. Vieira JM, Schwarz Q, Ruhrberg C. Selective requirements for NRP1 ligands during neurovascular patterning. Development 2007; 134(10):1833-1843.

119. Fuh G, Garcia KC, de Vos AM. The interaction of neuropilin-1 with vascular endothelial growth factor and its receptor flt-1. J Biol Chem 2000; 275(35):26690-26695.

120. Whitaker GB, Limberg BJ, Rosenbaum JS. Vascular endothelial growth factor receptor-2 and neuropilin-1 form a receptor complex that is responsible for the differential signaling potency of VEGF(165) and VEGF(121). J Biol Chem 2001; 276(27):25520-25531.

121. Soker S, Miao HQ, Nomi M et al. VEGF165 mediates formation of complexes containing VEGFR-2 and neuropilin-1 that enhance VEGF165-receptor binding. J Cell Biochem 2002; 85(2):357-368.

122. Wang L, Zeng H, Wang P et al. Neuropilin-1-mediated vascular permeability factor/vascular endothelial growth factor-dependent endothelial cell migration. J Biol Chem 2003; 278(49):48848-48860.

123. Cai H, Reed RR. Cloning and characterization of neuropilin-1-interacting protein: A PSD-95/Dlg/ ZO-1 domain-containing protein that interacts with the cytoplasmic domain of neuropilin-1. J Neurosci 1999; 19(15):6519-6527.

124. Chittenden TW, Claes F, Lanahan AA et al. Selective regulation of arterial branching morphogenesis by synectin. Dev Cell 2006; 10(6):783-795.

125. Mukouyama YS, Gerber HP, Ferrara N et al. Peripheral nerve-derived VEGF promotes arterial differentiation via neuropilin 1-mediated positive feedback. Development 2005; 132(5):941-952.

126. Stalmans I, Ng YS, Rohan R et al. Arteriolar and venular patterning in retinas of mice selectively expressing VEGF isoforms. J Clin Invest 2002; 109(3):327-336.

127. Bernfield M, Kokenyesi R, Kato M et al. Biology of the syndecans: A family of transmembrane heparan sulfate proteoglycans. Annu Rev Cell Biol 1992; 8:365-393.

128. Park JE, Keller GA, Ferrara N. The vascular endothelial growth factor (VEGF) isoforms: Differential deposition into the subepithelial extracellular matrix and bioactivity of extracellular matrix-bound VEGF. Mol Biol Cell 1993; 4(12):1317-1326.

129. Gerhardt H, Golding M, Fruttiger M et al. VEGF guides angiogenic sprouting utilizing endothelial tip cell filopodia. J Cell Biol 2003; 161(6):1163-1177.

130. Tessler S, Rockwell P, Hicklin D et al. Heparin modulates the interaction of VEGF165 with soluble and cell associated flk-1 receptors. J Biol Chem 1994; 269(17):12456-12461.

131. Terman B, Khandke L, Dougher-Vermazan M et al. VEGF receptor subtypes KDR and FLT1 show different sensitivities to heparin and placenta growth factor. Growth Factors 1994; 11(3):187-195.

132. Ashikari-Hada S, Habuchi H, Kariya Y et al. Heparin regulates vascular endothelial growth factor165-dependent mitogenic activity, tube formation, and its receptor phosphorylation of human endothelial cells: Comparison of the effects of heparin and modified heparins. J Biol Chem 2005; 280(36):31508-31515.

133. Jakobsson L, Kreuger J, Holmborn K et al. Heparan sulfate in trans potentiates VEGFR-mediated angiogenesis. Dev Cell 2006; 10(5):625-634.

134. Shintani Y, Takashima S, Asano Y et al. Glycosaminoglycan modification of neuropilin-1 modulates VEGFR2 signaling. Embo J 2006; 25(13):3045-3055.

135. Dougher M, Terman BI. Autophosphorylation of KDR in the kinase domain is required for maximal VEGF-stimulated kinase activity and receptor internalization. Oncogene 1999; 18(8):1619-1627.

136. Feng Y, Venema VJ, Venema RC et al. VEGF induces nuclear translocation of Flk-1/KDR, endothelial nitric oxide synthase, and caveolin-1 in vascular endothelial cells. Biochem Biophys Res Commun 1999; 256(1):192-197.

137. Li W, Keller G. VEGF nuclear accumulation correlates with phenotypical changes in endothelial cells. J Cell Sci 2000; 113(Pt 9):1525-1534.

138. Ilan N, Tucker A, Madri JA. Vascular endothelial growth factor expression, beta-catenin tyrosine phosphorylation, and endothelial proliferative behavior: A pathway for transformation? Lab Invest 2003; 83(8):1105-1115.

139. Santos SC, Miguel C, Domingues I et al. VEGF and VEGFR-2 (KDR) internalization is required for endothelial recovery during wound healing. Exp Cell Res 2007; 313(8):1561-1574.

140. Pan Q, Chanthery Y, Liang WC et al. Blocking neuropilin-1 function has an additive effect with anti-VEGF to inhibit tumor growth. Cancer Cell 2007; 11(1):53-67.

141. Hamada K, Sasaki T, Koni PA et al. The PTEN/PI3K pathway governs normal vascular development and tumor angiogenesis. Genes Dev 2005; 19(17):2054-2065.

CHAPTER 3

VEGF Gene Regulation

Marcus Fruttiger*

Abstract

VEGF is best known for its angiogenic properties. Not only does it promote the growth of new blood vessels during embryonic development, it is also important in the adult, where it plays a role in maintaining an adequate supply of oxygen and nutrients to most tissues. *VEGF* gene regulation is controlled by different signalling pathways depending on the context in which it is expressed. Best understood is the induction of VEGF expression by hypoxia in neonates and adults, which represents an adaptive response to metabolic stress. In contrast, the mechanisms that control VEGF expression during embryonic development are currently less clear.

Key Messages

- VEGF is a multifunctional molecule that is regulated by numerous different signalling pathways.
- VEGF expression is induced by hypoxia.
- Hypoxia stimulates *VEGF* transcription by increasing HIF activity.
- Hypoxia stimulates *VEGF* translation by increasing mRNA stability.
- *VEGF* gene expression is also controlled by hypoxia-independent mechanisms, in particular during embryogenesis.

Introduction

More than half a century ago, Michaelson used an ink perfusion technique to visualize the developing retinal vasculature and noticed that capillary growth near veins was much more vigorous than near arteries.[1] He proposed the existence of a vasoformative molecule termed factor X, which is (a) produced by extra-vascular tissue, (b) distributed in a gradient and (c) antagonized by oxygen. As we now know, these criteria are fulfilled by the vascular endothelial growth factor VEGF, also known as VEGFA. A team lead by Eli Keshet was the first to propose that VEGF could be the long elusive factor X that mediates hypoxia-induced vascular growth.[2] This was based on the knowledge that VEGF had already been shown to promote the growth of blood vessels,[3] and on the observation that VEGF mRNA dramatically increases under hypoxic conditions in various cell lines. In addition, it was found that VEGF levels were increased in the hypoxic centre of tumours, suggesting that this factor could mediate the growth of new vessels into tumours. This had important clinical implications, because it was known since the early seventies that tumour growth requires sprouting of new vessels from preexisting vessels, a process known as angiogenesis.[4] The discovery of an angiogenic factor in tumours provided for the first time a molecular target for anti-angiogenesis therapy in the fight against cancer,[5] triggering massive research on VEGF. In 1996 two teams simultaneously reported the

*Marcus Fruttiger—UCL Institute of Ophthalmology, University College London, 11-43 Bath Street, London EC1V 9EL, UK. Email: m.fruttiger@ucl.ac.uk

VEGF in Development, edited by Christiana Ruhrberg. ©2008 Landes Bioscience and Springer Science+Business Media.

genetic deletion of *VEGF* in mice.[6,7] Both groups found that the inactivation of just one *VEGF* allele caused early embryonic lethality. This was unexpected and made further experiments technically challenging, as no heterozygous founder mice could be created. It also dramatically illustrated the importance of correct VEGF dosage during embryogenesis. Subsequent research has uncovered a multitude of mechanisms that tightly control VEGF dosage and its biological activity.

VEGF Gene Regulation through the Hypoxia Response Element

The most prominent stimulus for VEGF expression is hypoxia. In hypoxic tissue, VEGF is upregulated, and this stimulates blood vessel growth. Increased blood supply then alleviates the hypoxia, turning VEGF expression off again. This simple negative feedback loop ensures that supply and demand of oxygen in tissue are always adequately matched. But how does this process work at a molecular level? A short sequence in the 5' flanking region of the *VEGF* gene is important for VEGF induction by hypoxia.[8] This sequence element, termed hypoxia response element (HRE), was initially discovered as an enhancer element within the erythropoietin (*EPO*) gene.[9,10] The element is also present in many other hypoxia inducible genes and is a binding site for the transcription factor hypoxia-inducible factor 1 (HIF1). HIF1 is a heterodimer consisting of an alpha and beta subunit, both of which are basic helix-loop-helix PAS domain proteins.[11] The beta subunit is known as the aryl hydrocarbon receptor nuclear translocator (ARNT) and is constitutively expressed, whereas the alpha subunit, HIF1A, is regulated by hypoxia. Although the mRNA encoding HIF1A increases in hypoxic cells, the dominant mechanism of HIF1A regulation is post-translational (Fig. 1). Under normoxic

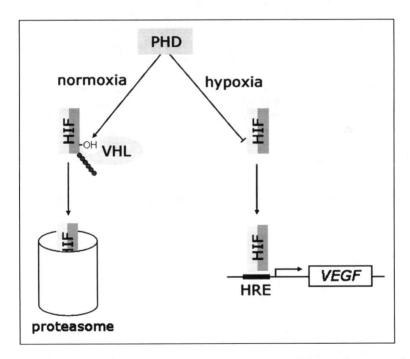

Figure 1. HIF mediated *VEGF* transcription. Under normoxic conditions, hypoxia inducible factor (HIF) is hydroxylated by prolyl hydroxylases (PHDs). This modification facilitates binding of the von Hippel Lindau protein (VHL), resulting in ubiquitination and rapid degradation of HIF. At low oxygen concentration, hydroxylation via PHDs becomes less efficient, resulting in HIF accumulation and HIF binding to a hypoxia response element (HRE) in the *VEGF* promoter.

conditions, HIF1A is swiftly destroyed via proteasomal degradation. However, when oxygen tension falls, this degradation process becomes less efficient and HIF1A protein rapidly accumulates; a sensible strategy, considering that under stressful conditions it might be easier to stop destroying a protein rather than to start producing new protein.

The gene product of the von-Hippel-Lindau (VHL) tumour suppressor gene is critically involved in the degradation of HIF1A. VHL is part of a multi protein complex containing ubiquitin E3 ligase. Binding of VHL to HIF1A results in HIF1A ubiquitination and very rapid proteasomal destruction, making HIF1A one of the most short-lived proteins known. In cells that lack VHL, HIF1A protein levels build up, resulting in increased expression of VEGF and other angiogenic factors. Clinical manifestation of this occurs in von Hippel-Lindau disease, a dominantly inherited familial cancer syndrome predisposing the patient to a variety of malignant and benign tumours. Germline mutations in one allele of VHL do not seem to have any noticeable deleterious effects. However, silencing the second copy of the gene via somatic inactivation or deleterious mutations frequently leads to haemangioblastomas and renal cell carcinomas.

The molecular mechanisms that control binding of VHL to HIF1A are broadly understood and depend on the enzymatic hydroxylation of conserved prolyl residues within the oxygen-dependent degradation domain (ODD) of HIF1A. Deletion of the ODD domain, or mutation of the prolyl residues to alanins, renders HIF resistant to oxygen-induced degradation. The hydroxylation of HIF1A can be carried out by at least three different prolyl hydroxylases, termed PHD1, PHD2 and PHD3 (new names: EGLN1-3). This modification requires the cofactors 2-oxoglutarate, vitamin C, iron and, importantly, molecular oxygen. At low oxygen concentrations, the PHDs become less efficient and HIF1A is no longer hydroxylated, which prevents HIF1A from binding VHL and therefore stops its ubiquitination and degradation; the resulting increased HIF levels ultimately augment VEGF transcription.

The pathway shown in Figure 1 is probably the best know mechanism of VEGF regulation to date. However, when the HRE in the *VEGF* promoter was deleted in mice, they showed an unexpected phenotype.[12] Consistent with an important role for HIF mediated VEGF expression during development, half of the mutant mice died embryonically or perinatally. The surviving half of mice lived for up to two years, but gained less weight, were infertile and developed adult-onset motor neuron degeneration resembling amyotrophic lateral sclerosis (ALS). Interestingly, these surviving mice show impaired hypoxia-induced VEGF upregulation in the brain and spinal cord, but not in fibroblasts, confirming the previous finding that the response to hypoxia is differentially regulated in an organ- and even cell type-specific manner.[13] However, these in vivo findings contrast a number of in vitro experiments, which had convincingly shown that hypoxic VEGF induction critically depends on HIF and the HRE in many different cell types.[8] It is possible that HIF influences *VEGF* gene expression via elements other than the HRE in the *VEGF* promoter; alternatively, HIF-independent systems may compensate for deficiency in the HRE-mediated hypoxia regulation in some tissues.

Post-Transcriptional Regulation of VEGF

The mRNAs of house keeping genes such as β-globin or GAPDH have half-lives of over 20 hours, whereas the mRNAs of cytokines or transcriptional activators typically are short lived, with half-lives of 10-30 minutes. Stabilization of normally unstable mRNAs in response to stimuli such as hypoxia, growth factors, hormones or second messengers provides a mechanism to regulate protein levels. Accordingly, Levy et al found that hypoxia increases VEGF mRNA half-life by a factor of 3, resulting in 8-30 times higher levels of VEGF mRNA. In comparison, HIF mediated regulation of VEGF transcription increases VEGF mRNA amounts only 2-3 times.[14]

Degradation of mRNA transcripts usually starts with 3'-5' exonucleolytic deadenylation, which removes most or all of the poly(A)-tail. The rest of the mRNA is degraded either by 3'-5' exonucleolytic degradation and/or by removal of the 5' cap, followed by 5'-3' exonucleolytic degradation.[15] Some mRNAs such as that for insulin-like growth factor are also cleaved by endonucleases, initiating subsequent exonucleolytic degradation.[16] Whilst the proteins that mediate and control this degradation process are only partially understood, a variety of mRNA cis-elements have been identified, which are decisive in determining mRNA stability. In the case of the VEGF mRNA, AU-rich cis-elements (AREs) are important. They are around 50 to 150 nucleotides long with no single conserved consensus motive, but they often contain several copies of the pentamer AUUUA or the nonamer UUAUUUAUU and one or several U-rich stretches. The 3' untranslated region of VEGF mRNA contains a 125bp ARE that is bound by a series of hypoxia-induced proteins,[17] but few have so far been identified. One of the binding partners of the VEGF-ARE is HuR (also known as ELAVL1), a member of the ELAV/Hu family of RNA binding proteins. HuR can stabilize mRNA by displacing or inhibiting factors that cleave or deadenylate ARE-containing transcripts.[18] Interestingly, brain tumours ubiquitously express HuR and display elevated levels of cytokines and angiogenic factors such as VEGF.[19] Poly(A)-binding protein interacting protein 2 (PAIP2) also binds the VEGF-ARE and, via interactions with HuR, stabilizes the VEGF mRNA.[20] A further VEGF-ARE binding protein is zinc finger binding protein 36 (ZFP36L1, also known as TIS11B), but, in contrast to HuR, it has mRNA-destabilizing activity.[21]

How hypoxia influences the activity of HuR or other VEGF mRNA-binding proteins remains unclear. An obvious possibility is that the same oxygen sensing machinery that regulates HIF stability feeds into mRNA-degradation mechanisms via interactions with VHL. In fact, there are studies that support such a view. For example, overexpression of VHL in renal carcinoma cells can decrease VEGF mRNA levels via a posttranscriptional mechanism,[22,23] and rapid turnover of mRNA containing AREs depends on ubiquitination and proteasome activity.[24] Furthermore, an association between HuR and VHL was shown in renal cell carcinoma, implicating VHL not only in HIF-mediated VEGF transcription, but also in the cellular signalling events that regulate VEGF mRNA stability.[25] On the other hand, VEGF mRNA can be stabilized by the protein kinase C signalling pathway[26] or the stress signalling pathways via p38MAPK and JNK.[27] Whether these pathways are used to mediate hypoxia-induced VEGF mRNA stabilization is not known.

In addition to mRNA stabilization, the post-transcriptional regulation of VEGF has one more layer of complexity that is based on mRNA translation. Eukaryotic protein synthesis usually depends on binding of the translation initiation complex to the mRNA cap,[28] with ribosomes scanning the 5' untranslated region (UTR) until they encounter the first AUG codon. VEGF possesses an unusually long GC-rich 5' UTR of more than 1000 nucleotides, which can inhibit efficient ribosomal scanning.[29] An alternative, cap-independent mechanism of translation has been identified in picornavirus, which contains long 5' UTRs without caps, but uses elements termed internal ribosomal entry sites (IRES). It is believed that VEGF mRNA with its cumbersome 5' UTR is usually translated efficiently due to the presence of an IRES element in the 5' UTR.[30] The cap-independent translation of VEGF mRNA via the IRES element remains efficient under hypoxic stress, whilst overall cap-dependent protein synthesis is reduced by up to 50%,[30] resulting in a relative increase of VEGF translation.

VEGF Gene Regulation by Transcription Factors

The signalling pathways that regulate cell metabolism and cell growth are intricately linked with the hypoxia-sensing machinery. This is illustrated by the fact that HIF induces angiogenic genes, but also regulates many aspects of anaerobic metabolism, such as the expression of glucose transporter 1 and most of the glycolytic enzymes.[31] VEGF plays a major role in helping to maintain cellular homeostasis in response to hypoxia, acidosis, UV, lack of nutrients and other stresses by stimulating angiogenesis. Accordingly, VEGF expression is not only controlled by hypoxia, but by a complex regulatory network involving many other signalling pathways, including cytokines and growth factors.[32] That the regulation of cell metabolism and the hypoxia-sensing machinery are interconnected is also evidenced in the control of *VEGF* gene expression: VEGF is rapidly induced upon serum stimulation in normoxic serum-deprived fibroblasts; this response is mediated by the ERK signalling pathway.[33] Activation of this pathway stimulates HIF activity as well,[34] which in turn also induces VEGF transcription.

There are several HIF-independent mechanisms to upregulate VEGF transcription. This is evident from the promoter region of the *VEGF* gene, which contains binding elements for transcription factors such as specific protein 1 (SP1), signal transducer and activator of transcription 3 (STAT3), activator protein 1 (AP1) and many others.[32,35] SP1 is essential for basal transcription of VEGF.[36] Its phosphorylation following ERK activation leads to VEGF mRNA upregulation independent of hypoxia and HIF.[33] It was also shown that SP1 is responsible for the very high levels of constitutively expressed VEGF in many cancer cell lines.[37] Furthermore, tumour suppressors such as p73 or VHL can affect VEGF transcription via SP1.[38,39] In the case of VHL, this is based on direct binding to SP1, which inhibits SP1 activity.[38] Thus, loss of VHL not only induces VEGF via HIF activation (see above), but also via de-repression of SP1. STAT3 is able to directly bind the *VEGF* promoter to upregulate transcription, and a constitutively active mutant form of STAT3 has been found in various types of tumours.[40,41] The presence of four candidate AP1 binding sites in the *VEGF* promoter suggests that AP1 plays a role in VEGF regulation. The AP1 transcription factor is a dimer consisting of basic region-leucine zipper (bZIP) proteins such as Jun and Fos protein family members. It can cooperate with HIF to increase VEGF expression under hypoxic conditions, which involves members of the MAP kinase family.[42] It has also been shown that JUND protects cells from oxidative stress and exerts an anti-angiogenic effect via a HIF-mediated mechanism.[43]

Transcription factors that directly regulate VEGF transcription, such as HIF, SP1, STAT3 and AP1, are in turn controlled by a complex network of interacting signalling pathways. Apart from the PHD-VHL pathway, the ERK, JNK and p38 MAP kinase pathways can all modulate VEGF expression; moreover, signalling via PI3K, AKT (also known as RAC) and mTOR (also known as FRAP1) plays a role in *VEGF* gene regulation.[32,35] Therefore, the simple concept that hypoxia controls HIF to control VEGF levels had to be broadened to include the idea that VEGF transcription and VEGF translation are being influenced by several other signalling pathways (Fig. 2).

Developmental Control of VEGF Expression

In all animals, small blood vessels and capillaries are organized stochastically, whereas larger vessels form stereotypical vascular trees. This makes sense, as small vessels grow in response to metabolic demand. On the other hand, large vessels look more or less the same in every person, because they were patterned by "hard wired" morphogenetic programs operating during embryonic development. VEGF plays a major role in vessel formation both in physiological and pathological angiogenesis in the adult as well as in the developing embryo. We now know that angiogenesis in the adult is to a large extent hypoxia-driven. But is developmental VEGF expression regulated by hypoxia or hypoxia-independent mechanisms? There is evidence supporting both mechanisms of VEGF-stimulated angiogenesis during development.

The neonatal retina is a particularly illustrative example of a tissue in which hypoxia plays a major role in VEGF regulation and angiogenesis during development. Around birth in mice and before birth in humans, the retinal vasculature emerges from the optic nerve head and covers the inner retinal surface as a centrifugally spreading vascular plexus. This process is tightly controlled by retinal astrocytes, which also emerge from the optic nerve head and migrate ahead of the growing blood vessel (Fig. 3). These specialized glial cells provide a template for the vascular network and strongly express VEGF. However, as soon as they become covered by the oxygen-carrying vessel network, VEGF is downregulated (Fig. 3). This differential expression of VEGF (low in the centre and high in the periphery of the retina) leads to a VEGF gradient across the retina that is important for the correct outgrowth of the retinal vasculature.[44] Labelling with a hypoxia-probe shows that the peripheral, not yet vascularized region of the retina is indeed hypoxic, whereas the vascularized central portion is not (Fig. 3). This

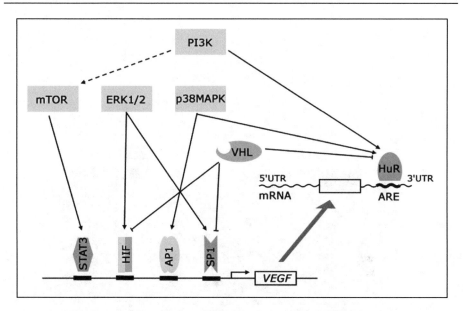

Figure 2. VEGF expression is regulated at the level of mRNA transcription and mRNA stability. VEGF transcription is induced by binding of diverse transcription factors to their recognition elements in the *VEGF* promoter. *VEGF* mRNA stability is increased by mRNA stabilizing proteins such as HuR (ELAVL1). A complex network of signalling pathways including different kinase pathways and VHL can influence both mRNA transcription and mRNA stability of *VEGF*.

difference in tissue oxygenation explains the high levels of VEGF mRNA in the periphery and the low levels in the central vascularized portion of the retina.[45]

It seems that the hypoxia-driven *VEGF* gene expression of peripheral retinal astrocytes does not depend on the HRE in the *VEGF* promoter, because the retinal vasculature develops normally in mice that lack this sequence in their genome.[46] It is therefore likely that the high levels of VEGF mRNA in peripheral retinal astrocytes are due to increased mRNA stability. Furthermore, retinal astrocytes express much higher levels of VEGF than the underlying neurons, even though both cell populations are likely to experience more or less the same oxygen tension. Retinal astrocytes may therefore be more sensitive to hypoxia than neurons; one could envisage a mechanism involving increased basal activity of the *VEGF* promoter, caused, for example, by increased SP1 levels in retinal astrocytes, but this has not yet been investigated.

It is conceivable that in the embryo, similar to the neonatal mouse retina, rapidly expanding tissue outgrows its vascular support and becomes hypoxic, inducing VEGF expression. Consistent with this idea, the use of hypoxia probes has revealed that the brain is hypoxic during embryogenesis.[47] In some brain regions, hypoxia-probed labelled areas correspond to areas prominent in HIF1A—and VEGF-immunoreactivity.[48] However, it is unlikely that a hypoxia-controlled mechanism is sufficient to pattern the entire embryonic vasculature. For example, VEGF plays a crucial role very early during development to promote the propagation of the endothelial precursor lineage and in the formation of the first major vessels, the dorsal aorta and the cardinal veins. Yet, these processes occur at a stage when the embryo is still small enough for oxygenation to occur via diffusion from the vascularised yolk sac. In these situations, VEGF is expressed in stereotypical and discrete areas to stimulate reproducible vascular patterning events. These expression patterns are likely "hard-wired", i.e., controlled by hypoxia-independent pathways. Consistent with this notion, sonic hedgehog is secreted from the notochord in zebrafish to induce VEGF expression in adjacent somites,

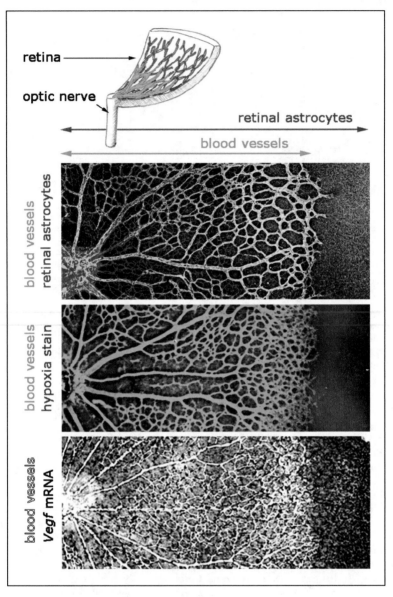

Figure 3. *VEGF* expression during retinal vascularisation. Astrocytes and blood vessels invade the newborn retina from the optic nerve (A). Retinal astrocytes serve as a template and VEGF source for the growing retinal vasculature (B). Astrocytes in the peripheral, not yet vascularized part of the retina experience hypoxia (C) and express *VEGF* at higher levels than astrocytes in the vascularized portion of the retina (D). The resulting VEGF gradient stimulates vessel growth towards the periphery.

which in turn leads to the formation of the dorsal aorta.[49] Moreover, during brainstem development, VEGF is expressed in stereotypical longitudinal stripes reminiscent of neurogenesis pattern well after the brain has assembled a vascular network to counteract the emergence of hypoxia; these expression patterns likely reflect a role in the development of neurons (and glia), rather than a role in blood vessel growth.[50] The signalling pathways that control VEGF expression during vasculogenesis are only beginning to be understood (see Chapter by L.C5. Goldie, M.K. Nix and K.K. Hirschi), and the pathways controlling VEGF expression during neuronal development are not known at all.

Conclusions

VEGF is a versatile molecule with different functions in different settings. It acts as a survival factor, mitogen and guidance molecule for endothelial cells. It regulates angiogenesis in response to metabolic needs in the adult, but also participates in the formation of the vascular tree during embryogenesis. In addition, VEGF can regulate vessel permeability and plays a role in neuronal development and neuroprotection.[50-52] With such a varied array of biological functions it is not surprising that the regulation of *VEGF* gene expression is diverse. During development, hypoxia-independent pathways likely control VEGF expression to induce the formation of the stereotypically patterned elements of the vascular tree, whereas hypoxia-induced VEGF expression can fine tune the shape and density of capillary networks. Importantly, *VEGF* gene expression is controlled by a complex network of intersecting signalling pathways containing several dozen growth factors and cytokines, many of which induce VEGF expression in cancer cells.[53] Future anti-angiogenic cancer therapies attempting to manipulate VEGF expression will have to take this complexity into account, as cancer cells may achieve VEGF upregulation both through HIF-related mechanisms as well as hypoxia-independent pathways, which are normally used to control VEGF expression during development. The hardwired mechanisms of VEGF regulation are poorly understood and pose a major challenge for future research.

References

1. Michaelson IC. Retinal circulation in man and animals. Springfield, IL: Charles C. Thomas, 1954.
2. Shweiki D, Itin A, Soffer D et al. Vascular endothelial growth factor induced by hypoxia may mediate hypoxia-initiated angiogenesis. Nature 1992; 359(6398):843-845.
3. Leung DW, Cachianes G, Kuang WJ et al. Vascular endothelial growth factor is a secreted angiogenic mitogen. Science 1989; 246(4935):1306-1309.
4. Folkman J. Tumor angiogenesis: Therapeutic implications. N Engl J Med 1971; 285(21):1182-1186.
5. Ferrara N. VEGF as a therapeutic target in cancer. Oncology 2005; 69(Suppl 3):11-16.
6. Carmeliet P, Ferreira V, Breier G et al. Abnormal blood vessel development and lethality in embryos lacking a single VEGF allele. Nature 1996; 380(6573):435-439.
7. Ferrara N, Carver-Moore K, Chen H et al. Heterozygous embryonic lethality induced by targeted inactivation of the VEGF gene. Nature 1996; 380(6573):439-442.
8. Forsythe JA, Jiang BH, Iyer NV et al. Activation of vascular endothelial growth factor gene transcription by hypoxia-inducible factor 1. Mol Cell Biol 1996; 16(9):4604-4613.
9. Madan A, Curtin PT. A 24-base-pair sequence 3' to the human erythropoietin gene contains a hypoxia-responsive transcriptional enhancer. Proc Natl Acad Sci USA 1993; 90(9):3928-3932.
10. Wang GL, Semenza GL. General involvement of hypoxia-inducible factor 1 in transcriptional response to hypoxia. Proc Natl Acad Sci USA 1993; 90(9):4304-4308.
11. Wang GL, Jiang BH, Rue EA et al. Hypoxia-inducible factor 1 is a basic-helix-loop-helix-PAS heterodimer regulated by cellular O2 tension. Proc Natl Acad Sci USA 1995; 92(12):5510-5514.
12. Oosthuyse B, Moons L, Storkebaum E et al. Deletion of the hypoxia-response element in the vascular endothelial growth factor promoter causes motor neuron degeneration. Nat Genet 2001; 28(2):131-138.
13. Marti HH, Risau W. Systemic hypoxia changes the organ-specific distribution of vascular endothelial growth factor and its receptors. Proc Natl Acad Sci USA 1998; 95(26):15809-15814.
14. Levy AP. Hypoxic regulation of VEGF mRNA stability by RNA-binding proteins. Trends Cardiovasc Med 1998; 8(6):246-250.
15. Hollams EM, Giles KM, Thomson AM et al. MRNA stability and the control of gene expression: Implications for human disease. Neurochem Res 2002; 27(10):957-980.

16. Meinsma D, Scheper W, Holthuizen PE et al. Site-specific cleavage of IGF-II mRNAs requires sequence elements from two distinct regions of the IGF-II gene. Nucleic Acids Res 1992; 20(19):5003-5009.
17. Claffey KP, Shih SC, Mullen A et al. Identification of a human VPF/VEGF 3' untranslated region mediating hypoxia-induced mRNA stability. Mol Biol Cell 1998; 9(2):469-481.
18. Zhao Z, Chang FC, Furneaux HM. The identification of an endonuclease that cleaves within an HuR binding site in mRNA. Nucleic Acids Res 2000; 28(14):2695-2701.
19. Nabors LB, Gillespie GY, Harkins L et al. HuR, a RNA stability factor, is expressed in malignant brain tumors and binds to adenine- and uridine-rich elements within the 3' untranslated regions of cytokine and angiogenic factor mRNAs. Cancer Res 2001; 61(5):2154-2161.
20. Onesto C, Berra E, Grepin R et al. Poly(A)-binding protein-interacting protein 2, a strong regulator of vascular endothelial growth factor mRNA. J Biol Chem 2004; 279(33):34217-34226.
21. Ciais D, Cherradi N, Bailly S et al. Destabilization of vascular endothelial growth factor mRNA by the zinc-finger protein TIS11b. Oncogene 2004; 23(53):8673-8680.
22. Gnarra JR, Zhou S, Merrill MJ et al. Post-transcriptional regulation of vascular endothelial growth factor mRNA by the product of the VHL tumor suppressor gene. Proc Natl Acad Sci USA 1996; 93(20):10589-10594.
23. Iliopoulos O, Levy AP, Jiang C et al. Negative regulation of hypoxia-inducible genes by the von Hippel-Lindau protein. Proc Natl Acad Sci USA 1996; 93(20):10595-10599.
24. Laroia G, Sarkar B, Schneider RJ. Ubiquitin-dependent mechanism regulates rapid turnover of AU-rich cytokine mRNAs. Proc Natl Acad Sci USA 2002; 99(4):1842-1846.
25. Datta K, Mondal S, Sinha S et al. Role of elongin-binding domain of von Hippel Lindau gene product on HuR-mediated VPF/VEGF mRNA stability in renal cell carcinoma. Oncogene 2005; 24(53):7850-7858.
26. Shih SC, Mullen A, Abrams K et al. Role of protein kinase C isoforms in phorbol ester-induced vascular endothelial growth factor expression in human glioblastoma cells. J Biol Chem 1999; 274(22):15407-15414.
27. Pages G, Berra E, Milanini J et al. Stress-activated protein kinases (JNK and p38/HOG) are essential for vascular endothelial growth factor mRNA stability. J Biol Chem 2000; 275(34):26484-26491.
28. Pain VM. Initiation of protein synthesis in eukaryotic cells. Eur J Biochem 1996; 236(3):747-771.
29. Kozak M. Pushing the limits of the scanning mechanism for initiation of translation. Gene 2002; 299(1-2):1-34.
30. Stein I, Itin A, Einat P et al. Translation of vascular endothelial growth factor mRNA by internal ribosome entry: Implications for translation under hypoxia. Mol Cell Biol 1998; 18(6):3112-3119.
31. Semenza G. Signal transduction to hypoxia-inducible factor 1. Biochem Pharmacol 2002; 64(5-6):993-998.
32. Xie K, Wei D, Shi Q et al. Constitutive and inducible expression and regulation of vascular endothelial growth factor. Cytokine Growth Factor Rev 2004; 15(5):297-324.
33. Milanini J, Vinals F, Pouyssegur J et al. p42/p44 MAP kinase module plays a key role in the transcriptional regulation of the vascular endothelial growth factor gene in fibroblasts. J Biol Chem 1998; 273(29):18165-18172.
34. Berra E, Milanini J, Richard DE et al. Signaling angiogenesis via p42/p44 MAP kinase and hypoxia. Biochem Pharmacol 2000; 60(8):1171-1178.
35. Pages G, Pouyssegur J. Transcriptional regulation of the Vascular Endothelial Growth Factor gene—a concert of activating factors. Cardiovasc Res 2005; 65(3):564-573.
36. Ryuto M, Ono M, Izumi H et al. Induction of vascular endothelial growth factor by tumor necrosis factor alpha in human glioma cells: Possible roles of SP-1. J Biol Chem 1996; 271(45):28220-28228.
37. Shi Q, Le X, Abbruzzese JL et al. Constitutive Sp1 activity is essential for differential constitutive expression of vascular endothelial growth factor in human pancreatic adenocarcinoma. Cancer Res 2001; 61(10):4143-4154.
38. Mukhopadhyay D, Knebelmann B, Cohen HT et al. The von Hippel-Lindau tumor suppressor gene product interacts with Sp1 to repress vascular endothelial growth factor promoter activity. Mol Cell Biol 1997; 17(9):5629-5639.
39. Salimath B, Marme D, Finkenzeller G. Expression of the vascular endothelial growth factor gene is inhibited by p73. Oncogene 2000; 19(31):3470-3476.
40. Niu G, Wright KL, Huang M et al. Constitutive Stat3 activity up-regulates VEGF expression and tumor angiogenesis. Oncogene 2002; 21(13):2000-2008.
41. Wei D, Le X, Zheng L et al. Stat3 activation regulates the expression of vascular endothelial growth factor and human pancreatic cancer angiogenesis and metastasis. Oncogene 2003; 22(3):319-329.
42. Michiels C, Minet E, Michel G et al. HIF-1 and AP-1 cooperate to increase gene expression in hypoxia: Role of MAP kinases. IUBMB Life 2001; 52(1-2):49-53.

43. Gerald D, Berra E, Frapart YM et al. JunD reduces tumor angiogenesis by protecting cells from oxidative stress. Cell 2004; 118(6):781-794.
44. Gerhardt H, Golding M, Fruttiger M et al. VEGF guides angiogenic sprouting utilizing endothelial tip cell filopodia. J Cell Biol 2003; 161(6):1163-1177.
45. West H, Richardson WD, Fruttiger M. Stabilization of the retinal vascular network by reciprocal feedback between blood vessels and astrocytes. Development 2005; 132(8):1855-1862.
46. Vinores SA, Xiao WH, Aslam S et al. Implication of the hypoxia response element of the Vegf promoter in mouse models of retinal and choroidal neovascularization, but not retinal vascular development. J Cell Physiol 2006; 206(3):749-758.
47. Chen EY, Fujinaga M, Giaccia AJ. Hypoxic microenvironment within an embryo induces apoptosis and is essential for proper morphological development. Teratology 1999; 60(4):215-225.
48. Lee YM, Jeong CH, Koo SY et al. Determination of hypoxic region by hypoxia marker in developing mouse embryos in vivo: A possible signal for vessel development. Dev Dyn 2001; 220(2):175-186.
49. Lawson ND, Vogel AM, Weinstein BM. Sonic hedgehog and vascular endothelial growth factor act upstream of the Notch pathway during arterial endothelial differentiation. Dev Cell 2002; 3(1):127-136.
50. Schwarz Q, Gu C, Fujisawa H et al. Vascular endothelial growth factor controls neuronal migration and cooperates with Sema3A to pattern distinct compartments of the facial nerve. Genes Dev 2004; 18(22):2822-2834.
51. Rosenstein JM, Krum JM. New roles for VEGF in nervous tissue—beyond blood vessels. Exp Neurol 2004; 187(2):246-253.
52. Storkebaum E, Lambrechts D, Carmeliet P. VEGF: Once regarded as a specific angiogenic factor, now implicated in neuroprotection. Bioessays 2004; 26(9):943-954.
53. Loureiro RM, D'Amore PA. Transcriptional regulation of vascular endothelial growth factor in cancer. Cytokine Growth Factor Rev 2005; 16(1):77-89.

CHAPTER 4

Embryonic Vasculogenesis and Hematopoietic Specification

Lauren C. Goldie, Melissa K. Nix and Karen K. Hirschi*

Abstract

Vasculogenesis is the process by which blood vessels are formed de novo. In mammals, vasculogenesis occurs in parallel with hematopoiesis, the formation of blood cells. Thus, it is debated whether vascular endothelial cells and blood cells are derived from a common progenitor. Whether or not this is the case, there certainly is commonality among regulatory factors that control the differentiation and differentiated function of both cell lineages. VEGF is a major regulator of both cell types and plays a critical role, in coordination with other signaling pathways and transcriptional regulators, in controlling the differentiation and behavior of endothelial and blood cells during early embryonic development, as further discussed herein.

Key Messages

- Blood cells and blood vessel endothelium develop in parallel, but it is still debated whether they develop from a shared precursor.
- VEGF signaling is critical for vascular plexus formation and hematopoiesis.
- VEGF signaling cooperates with other signaling pathways to promote endothelial cell specification, vascular plexus remodeling and hematopoetic specification.
- VEGF signaling is fine tuned via a collection of receptors and co-receptors to modulate endothelial cell behavior.
- The precise contribution of different VEGF signaling pathways to the regulation of vascular development and hematopoiesis is not yet fully understood and needs to be researched further.

Introduction

Vasculogenesis is the process by which blood vessels are formed de novo. This process first occurs in the embryonic yolk sac of mammalian embryos, and then later during development in the embryo proper. During gastrulation, embryonic ectodermal (epiblast) cells are recruited to the primitive streak where they undergo an epithelial to mesenchymal transition. These cells then migrate between the visceral endoderm and epiblast to form either mesoderm or definitive endoderm.[1] In the yolk sac, the visceral endoderm is thought to elicit soluble signals which target the underlying mesoderm to induce the formation of primitive endothelial and hematopoietic cells, the first differentiated cell types to be produced in the mammalian embryo (Fig. 1). Primitive endothelial and hematopoietic cells coalesce to form blood islands that then fuse

*Corresponding Author: Karen K. Hirschi—Department of Pediatrics, Children's Nutrition Research Center and Department of Molecular and Cellular Biology, Center for Cell and Gene Therapy, Baylor College of Medicine, Houston, Texas 77030, USA. Email: khirschi@bcm.tmc.edu

VEGF in Development, edited by Christiana Ruhrberg. ©2008 Landes Bioscience and Springer Science+Business Media.

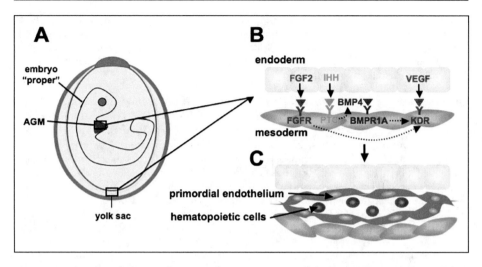

Figure 1. Molecular regulation of primitive hematopoietic specification and vascular remodeling during murine development. A) Schematic representation of the developing murine embryo, highlighting the primary sites of embryonic hematopoiesis. B) Schematic representation of endoderm-derived molecular signals with known roles in the specification of mesodermal precursors. C) Schematic representation of primitive hematopoietic development within the yolk sac vascular plexus.

to form a primitive network of tubules known as a capillary or vascular plexus. Remodeling and maturation of the capillary plexus into a circulatory network requires the subsequent recruitment of mural cells (smooth muscle cells and pericytes) to form the outer blood vessel wall. Concurrent with vascular remodeling is the induction of definitive hematopoiesis, the formation of mature circulating blood cells. This chapter focuses on understanding the sequential processes of vasculogenesis and hematopoiesis, and the major signaling pathways that regulate these processes; a key regulatory factor is vascular endothelial growth factor (VEGF), the main focus of this text.

Endothelial Specification and Vasculogenesis

During blood vessel development, endothelial cells are predominantly derived from mesoderm, as are blood cells. Mural cells are also largely derived from mesoderm, although neural crest and proepicardial organ cells also contribute mural cells to developing vessels. Vascular development has been well studied in several embryonic model systems including mouse, chick, quail, frog, and zebrafish, but perhaps best genetically manipulated in the murine system. Thus, our discussion of the molecular regulation of vasculogenesis and hematopoiesis will be based largely on insights gained from the study of the developing mouse; relevant insights from other model systems will also be discussed.

Formation of Mesodermal Precursors

Murine blood vessel formation initiates during gastrulation, wherein posterior epiblast cells migrate through the primitive streak to form mesoderm,[1] from which vascular and blood cells are differentiated. Several signals cooperate to promote mesoderm specification for blood vessel formation (Fig. 1B). One signal needed for specification of mesoderm is the soluble factor bone morphogenic protein 4 (BMP4), whose expression prior to gastrulation is localized to the primitive streak. In the absence of BMP4, mesoderm fails to develop and thus mutant embryos arrest at the egg chamber stage.[2] Another factor required for mesoderm formation is fibroblast growth factor 2 (FGF2, previously known as basic FGF or bFGF). Knockout inactivation of

FGF receptor 1 (FGFR1) results in an accumulation of epiblast cells, as they fail to migrate through the primitive streak and form mesoderm, further emphasizing the importance of FGF signaling in mesoderm formation.[3] VEGF has critical roles in the subsequent stages of mesodermal commitment and differentiation, but unlike BMP4 and FGF2, VEGF is not known to play a role in mesoderm formation.

Commitment of Mesodermal Precursors to the Endothelial Lineage

During embryogenesis, commitment of multipotent mesodermal cells to an endothelial cell lineage is thought to be regulated by soluble signals derived from adjacent endodermal cells. In the murine yolk sac, visceral endoderm elicits signals needed for vascular induction, although the hierarchy of such signals is not yet clear. One such soluble effector is indian hedgehog (IHH) (Fig. 1B). Using mouse embryo explant cultures lacking endoderm, it was discovered that IHH can respecify anterior epiblast cells, which normally form neuroectoderm, to form endothelial and hematopoietic cells instead,[4] thus demonstrating sufficiency of IHH for vascular induction. IHH's downstream effector in this pathway may be BMP4, as it is upregulated in response to IHH signaling (Fig. 1B).[4] In support of this theory, aberrant vasculogenesis and hematopoiesis has been demonstrated in BMP4-null mutant mouse embryos that survive past the egg cylinder chamber stage.[2]

Role of VEGF/KDR Signaling in Mesodermal Precursors of the Endothelial Lineage

Synergistic signaling between IHH and endoderm-derived FGF2 is thought to promote the expression of the VEGF receptor KDR (also known as FLK1 or VEGFR2) in adjacent mesodermal cells (Fig. 1B). A role for FGF2 in the formation of endothelial precursors, angioblasts, from mesoderm has been demonstrated in the quail embryo, where explant cultures consisting of only mesoderm lack the ability to undergo angioblast formation unless treated with FGF2.[5] Furthermore, isolated FGF2-treated epiblast cells are capable of forming endothelial cells in vitro.[6] Perhaps the critical role of FGF2 in vivo during vascular induction is the upregulation of KDR. Once KDR is upregulated in mesodermal progenitors (not all of which will become endothelial cells[7]), they become capable of responding to VEGF, which is initially produced only in the visceral endoderm of the mouse yolk sac.[8] KDR-expressing mesodermal progenitors are exquisitely sensitive to the bioactive levels of VEGF, which therefore exerts a dose-dependent effect on vasculogenesis. This is evidenced by the fact that mice heterozygous for a null mutation in the *Vegfa* gene die in utero due to failed vascular development: In these mutants, endothelial cells differentiate, but in a delayed fashion, and this leads to disorganized blood vessels and disrupts hematopoietic cell formation.[9] Similarly, embryos lacking KDR exhibit arrested vascular development and embryonic lethality. These mutants form endothelial precursors, but they fail to differentiate into mature endothelial cells, thereby halting blood island development and vascular plexus formation.[10] The data obtained from the KDR null mice have been corroborated in both the avian and *Xenopus* systems. In quail embryos, KDR expression was knocked down and revealed that primitive endothelial cells developed in the absence of KDR, but no mature endothelial networks formed.[11] In addition, it was demonstrated that quail mesoderm isolated from embryos can form hemangioblasts upon treatment with VEGF, even in the absence of contact with endoderm.[12] When VEGF was ectopically expressed in *Xenopus* embryos, large disorganized vascular structures containing mature endothelial cells formed.[13] Taken together, these data indicate that VEGF is critically important for the differentiation and/or survival of mature endothelial cells.

Role of VEGF/Neuropilin Signaling in the Endothelial Lineage

Recently, two novel VEGF receptors were discovered and termed neuropilin 1 and 2 (NRP1 and NRP2). Originally, the neuropilins were identified in the nervous system as transmembrane receptors for class III semaphorins, which provide repulsive cues during the axon guidance of several embryonic nerves.[14,15] Both NRP1 and NRP2 are also expressed on endothelial cells, where they function as isoform-specific VEGF receptors.[16,17] NRP1 null mice are embryonic lethal between E12.5 and E13.5 due to cardiovascular defects, including defective aortic arch remodeling and formation of abnormal capillary networks in the brain and spinal cord.[18] In contrast, NRP2 null mice have been reported to be viable and fertile.[19] However, ablation of both NRP1 and NRP2 results in embryonic lethality by E8.5 in the mouse, as no organized blood vessels are formed, similar to the KDR and VEGF null phenotypes.[20] This demonstrates, once again, the necessity of precise VEGF signaling for early vascular development.

Capillary Plexus Formation

As endothelial cells are committed and differentiated during vasculogenesis, their proliferation, migration and coalescence into a primitive capillary plexus must be well coordinated. VEGF signaling is critically important in this process, as it promotes endothelial cell proliferation[21-23] and modulates migration.[24-26] FGF2 is also involved in the stimulation of endothelial cell proliferation,[27] and has been shown to function synergistically with VEGF.[24,28,29] The mitogenic effects of VEGF are thought to be mediated largely by the KDR receptor.[26,30] In contrast, an alternative VEGF receptor termed FLT1 (also known as VEGFR1) is thought to modulate VEGF's proliferative effects by sequestering locally available soluble VEGF. Consistent with this idea is the phenotype of embryos lacking FLT1: These mutants exhibit excessive endothelial cell formation, which leads to the formation of a disorganized vascular plexus incapable of remodeling into a functional circulatory network.[31,32] In addition, a soluble form of FLT1 termed sFLT1 is thought to be critical to shape VEGF gradients in the environment of growing blood vessels to promote directional vessel growth.[33] These studies emphasize the fact that although VEGF's mitogenic effects are essential for capillary plexus formation, they must be balanced to prevent excessive endothelial cell proliferation and promote sprouting.

Vascular Remodeling

Once a primitive endothelium has been formed and patterned, the next aspect of blood vessel development is remodeling of the plexus to promote the establishment of a mature circulatory network.

Endothelial Cell Proliferation, Migration and Survival

Vascular remodeling is a complex process, in which a balance between signals to induce and inhibit endothelial cell proliferation must be reached. This process involves multiple signaling cascades as well as cell-cell and cell-matrix communications. Factors involved in maintaining the appropriate rate of endothelial cell proliferation include VEGF, FGF2, retinoic acid (RA), and transforming growth factor beta (TGFB1). VEGF and FGF2 are needed for the induction of endothelial cell proliferation. However counteractive anti-proliferative signals such as RA[34] and TGFB1[35] are an equally essential requirement for appropriate blood vessel formation.

Retinaldehyde dehydrogenase 2 (RALDH2) converts retinol (vitamin A) into its biologically active metabolite retinoic acid. Like other mesoderm modulating factors in the developing murine yolk sac, RALDH2 is expressed by the visceral endoderm, where active production of retinoic acid occurs. Retinoic acid is actively secreted from the visceral endoderm and interacts with retinoic acid receptors (RARA1/2) expressed by the vascular endothelium.[35] RARA1/2 signaling directly inhibits endothelial cell cycle progression via upregulation of two cell cycle inhibitors, the cyclin-dependent kinase inhibitors p21 (CDKN1A) and p27 (CDKN1B),[34] and indirectly suppresses growth via upregulation of the anti-proliferative cytokine TGFB1. TGFB1 likely acts through the SMAD5 pathway to upregulate the production of extracellular matrix protein fibronectin, which then functions to promote visceral endoderm survival and signals via integrins present on vascular endothelium. Of particular importance are integrins alpha 5 beta 1 and alpha V beta 3, which both bind fibronectin, but elicit opposite responses. Signalling of integrin alpha 5 beta 1 via a currently undefined intracellular signaling pathway serves to block endothelial cell proliferation; in contrast, integrin alpha v beta 3 signals via the KDR/MAPK pathway to promote endothelial cell proliferation.[35] The maintenance of a delicate equilibrium between these signaling events is necessary to regulate the proper branching and remodeling required for development of a mature vascular network.

Role of Mechanotransduction and Flow in the Maintenance of Blood Vessels

Once blood vessels have formed, and likely concurrent with the remodeling process, fortification of the vessel wall is initiated in response to hemodynamic forces, which are exerted upon endothelial cells by the newly established circulatory flow. Genomic studies in vitro have identified a multitude of genes that are differentially regulated by various types of hemodynamic forces, including laminar shear stress, turbulent shear stress, and disturbed flow.[36-40] These studies provide strong evidence that endothelial cells not only have the ability to "sense" hemodynamic forces, but that they are also capable of discriminating between different types of biomechanical stimuli. In particular, shear stress has been shown to promote endothelial cell survival via two main events, cell growth and inhibition of cell death (apoptosis). While the role of VEGF in endothelial response to shear stress is currently unknown, its receptor KDR appears of great importance: KDR is a mechanosensor that converts mechanical stimuli into chemical signals, as both its expression and activation level increase in response to laminar shear stress.[41] Upon exposure to fluid shear stress, endothelial cells also upregulate KDR expression to activate both ERK and JNK kinases, which are downstream targets of the MAPK pathway.[42] Once these pathways are initiated, the transcription of immediate early response genes such as monocyte chemotactic protein 1 and FOS is upregulated to promote endothelial cell growth.[43] Anti-apoptotic signaling pathways are also regulated by shear stress. Activation of the receptor tyrosine kinases KDR and TIE2 initiates a signaling cascade, in which activation of PI3K promotes the phosphorylation of AKT, which in turn triggers the upregulation of nitric oxide production and thereby elicits an antiapoptotic signal to endothelial cells in the blood vessel wall.[44-47] Endothelial expression of p21 drastically increases in response to increased nitric oxide production, resulting in the inhibition of endothelial cell apoptosis.[48] Vice versa, loss of p21 significantly increases endothelial cell death in response to shear stress, demonstrating that p21 is one of the major factors mediating the anti-apoptotic effect of shear stress.[48] In addition to its anti-apoptotic role, p21 creates a G1 to S phase block via cyclin-dependent kinase phosphorylation of the retinoblastoma protein,[49] thus decreasing the rate of DNA synthesis during exposure to laminar shear stress.

Hemogenic Specification and Hematopoiesis

Developmental Origin of Hematopoietic Progenitors

In the mouse, primitive hematopoiesis is initiated in the yolk sac between embryonic day 7.0 and 7.5, producing predominantly nucleated erythroid cells expressing embryonic globin.[50] A second wave of hematopoiesis is initiated in the yolk sac[51] and embryo proper[52] between E10.0 and E11.0,[53] when definitive hematopoietic stem cells (HSCs), capable of repopulating the neonate or adult blood system, arise in the aorta-gonad-mesonephros (AGM) region. Within the AGM, hematopoietic precursors are seen as clumps of cells that appear to bud from endothelial cells of the ventral wall of the dorsal aorta, and the umbilical and vitelline arteries.[54] Thus, during definitive as well as primitive yolk sac hematopoiesis, there is a close spatial and temporal association between endothelial and hematopoietic cell development. Studies in both chick and mouse have shown that AGM-derived yolk sac cells, which give rise to endothelial cells of the yolk sac vasculature, also have the capacity to generate definitive hematopoietic cells in vitro.[55] Together, these findings point to endothelium as the most likely source of definitive hematopoietic precursors in the developing embryo. However, the differing interpretation of the current body of experimental evidence has yielded three major theories for the origin and specification of the hematopoietic lineage (Fig. 2).

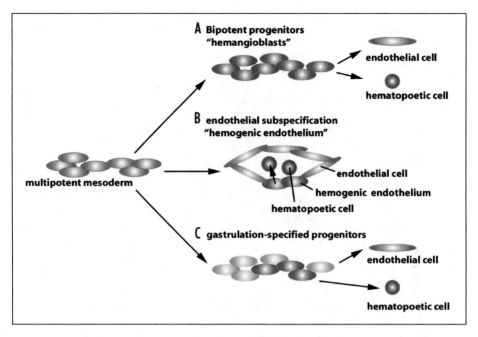

Figure 2. Theories of hematopoietic specification during embryogenesis. Schematic representation of the three major alternative theories for the developmental emergence of primitive endothelial and hematopoietic lineages: (A) the bipotent "hemangioblast", (B) "hemogenic" endothelium, and (C) gastrulation-specified endothelial and hematopoietic progenitors.

Endothelial and Hematopoietic Cells May Share a Common Progenitor

In blood islands of the yolk sac, where the earliest hematopoietic cells appear, cells in the hematopoietic and endothelial lineage arise almost simultaneously from the extraembryonic paraxial mesoderm to form a primitive capillary plexus, in which primitive nucleated erythroblasts are intimately associated with maturing endothelial cells.[56] This observation has led to the hypothesis that both lineages arise from a common precursor (Fig. 2A). This concept is supported by experimental evidence suggesting the shared expression of hematopoietic and endothelial-specific genes in both lineages. Even though the physical proximity of endothelial and hematopoietic cells in the developing embryo does not prove the existence of a common progenitor, supportive evidence for this concept is provided by in vitro studies, which demonstrate the formation of clonal blast colony-forming cells capable of generating both primitive hematopoietic and endothelial cell types under appropriate culture conditions.[57-59] These cells express markers associated with both hematopoietic and vascular development, including KDR,[60-62] TAL1,[50,63] VE-cadherin[64] and GATA1.[65] Cell tracking studies of transgenic murine embryos with GFP targeted to the *brachyury* locus have also identified clonal progenitors in the gastrulating mouse embryo that are capable of forming blast cell colonies with ex vivo hematopoietic and vascular potential.[66]

Hematopoietic Cells May Be Derived from Specialized Hemogenic Endothelium

An alternative theory, equally consistent with existing evidence, postulates that blood forming cells are specified from mesodermally-derived primitive endothelium known as *hemogenic endothelium* (Fig. 2B). Clusters of hematopoietic precursors that appear to derive directly from the vascular endothelium in vivo have been observed in developing blood vessels of numerous vertebrate species, and their appearance correlates with the developmental timing of definitive hematopoietic cells.[54] In the murine yolk sac and embryo proper, cells with blood-forming capacity have also been found to reside within the vascular endothelium.[67] A population of cells termed side population cells has been isolated from murine yolk sac and embryo prior to the onset of circulation; this cell population resides within the vascular endothelium, expresses high levels of endothelial markers, and fails to generate blood cells in vitro; however, as development progresses, this cell population acquires hematopoietic potential and phenotypic characteristics similar to those of bone marrow side population cells, suggesting a developmental transition from an endothelial to a hematopoietic progenitor cell type.[67] Culture studies also provided support for the model of hematopoietic specification from hemogenic endothelium. In particular, a population of primitive endothelial-like cells was isolated from human embryonic stem cell cultures, which expresses the endothelial markers PECAM1, KDR and VE-cadherin (PFV+), but not the hematopoetic lineage marker CD45; when treated with VEGF, these cells were shown to generate mature endothelial cells, whilst treatment with hematopoietic growth factors facilitated hematopoietic development.[68] These studies therefore identified PFV+/CD45- cells as a distinct bipotent progenitor population capable of supporting the progressive development of the endothelial and hematopoietic lineages. Most significantly, these studies demonstrated the stepwise generation of hematopoietic and endothelial cell precursors from multipotent embryonic cells, revealing a hierarchical organization of cell fate commitment, which supports the idea that hematopoietic and endothelial cells derive from a subset of embryonic endothelium with hemangioblastic properties.

Endothelial and Hematopoietic Progenitors May Be Independently Specified during Gastrulation

Other studies suggest that there is no common lineage between blood and endothelial cells, and that both are instead independently specified during gastrulation (Fig. 2C). Studies of the early primitive streak-stage embryo confirmed that cells derived from the primitive(streak preferentially contribute to the erythrocyte precursor population of the developing vascular blood islands.[1] In contrast, cells derived from the primitive streak of mid-primitive streak-stage embryos appear to contribute primarily to the endothelium, with little or no erythroid component.[1] Perhaps most significantly, simultaneous contribution of primitive streak progenitors to blood islands and vitelline endothelium is rarely seen to occur at the same location within yolk sac mesoderm.[1] These observations suggest that the allocation and recruitment of early progenitor cells to the endothelial and hematopoietic lineages likely occurs at different stages of gastrulation, and is therefore both spatially and temporally segregated during morphogenesis of the vitelline vessels.

Summary

The relationship between hematopoietic precursors derived from hemogenic endothelium and putative yolk sac hemangioblasts remains unclear. Progenitors capable of reconstituting neonatal hematopoiesis arise in the yolk sac before circulation,[51] and may be related to, or even derived from, the earlier born yolk sac hemangioblast. However, definitive hematopoietic stem cells that first arise in the AGM region are thought to be unrelated by lineage to yolk sac-derived precursors, because their development occurs independently of any circulation of cells from the yolk sac.[53] Isolation of single cells that can give rise to both endothelial and hematopoietic cells in vitro provides clear evidence of a common origin for the two lineages. However, it does not distinguish whether this common precursor makes a simple choice between the two lineages, or whether endothelium is formed first and blood forming cells are specified subsequently, or whether there is a more primitive multipotent precursor, whose potency remains undefined.

Developmental Regulation of Hematopoietic Progenitors

Although the cellular origin of hematopoietic cells is still debated, a number of signaling molecules have been shown to play critical roles in the specification and differentiation of blood cells.

VEGF/KDR

As previously discussed, KDR plays a key role in the developmental regulation of hematopoiesis as well as vasculogenesis. Generally recognized as the earliest antigenic marker of developing endothelial cells, *Kdr* expression can be detected in mesodermally derived blood island progenitors as early as E7.0.[62,69] Mice deficient in KDR fail to develop definitive hematopoietic cells capable of long-term engraftment in an irradiated recipient, nor do they make functional blood vessels, and they therefore die in midgestation, between E8.5 and 9.5.[10,70] In chimeric aggregation studies using wild-type mouse embryos, KDR null ES cells fail to contribute to primitive or definitive hematopoiesis in vivo, suggesting a cell autonomous requirement for KDR signaling in hematopoietic development.[70] Furthermore, VEGF signaling from the yolk sac endoderm via KDR is required for the hematopoietic differentiation of mesoderm.[9,32,70] Moreover, both vasculogenesis and hematopoiesis are impaired in mice homozygous for a hypomorphic VEGF allele, and decreased KDR activity.[8] Although the findings described above demonstrate that VEGF/KDR signalling plays a crucial role in promoting hematopoiesis, it may not be essential for the initial specification of hematopoietic precursors KDR-deficient ES cells retain some hematopoietic potential in vitro[70] and KDR null mouse because embryos contain normal numbers of hematopoietic progenitors at E7.5, even though they are profoundly deficient at E8.5.[71] Rather than being required for specification, VEGF/KDR signalling may provide an important signal for the subsequent survival, migration and clonal expansion of hematopoeic progenitors. Consistent with this idea, KDR+ cells of mouse embryos carrying a hypomorphic *Vegfa* allele reach the yolk sac on time by E8.5, but are severely compromised in their ability to generate primitive erythroid precursors, perhaps because VEGF normally prolongs the life span of primitive erythroid progenitors by inhibiting apoptosis.[72]

TAL1

The transcription factor TAL1 (also known as SCL) is first coexpressed with KDR around E7.0 in the visceral mesoderm and in vitro promotes the early stages of differentiation of mesoderm into cells of the hematopoietic lineage.[73,74] Gene targeting and chimera analyses demonstrated a requirement of TAL1 for the generation of both primitive and definitive hematopoietic lineages and for the appropriate remodeling of the yolk sac vasculature in vivo.[75-77] Specifically, TAL1 null mutant embryos are embryonic lethal, as they fail to initiate yolk sac hematopoiesis. KDR+ cells isolated from differentiating TAL1 null ES cells also fail to generate either blood or endothelial cells in vitro,[78] suggesting that KDR signaling and TAL1 expression synergise during early hematopoietic development.

GATA1, GATA2, LMO2 and RUNX1

The transcription factors GATA1, GATA2, LMO2 and RUNX1 also play important roles in the fate determination of the hematopoietic lineage. GATA1 and GATA2 are essential for both embryonic and adult erythropoiesis: Loss of GATA1 in knockout mouse embryos halts erythroid differentiation at the proerythroblast stage,[79] whilst loss of GATA2 function causes early embryonic lethality due to a primary defect in primitive hematopoiesis.[80] LMO2 is required for both vascular development and yolk sac hematopoiesis, and LMO2 null ES cells fail to contribute to the vascular endothelium of chimeric animals.[81,82] Loss of function of any one of these three transcription factors yields a phenotype similar to that of TAL1 null mutant mice, consistent with the idea that LMO2 may function cooperatively with GATA1 and TAL1 to promote the specification of erythroid cells.[81] Definitive, but not primitive, hematopoiesis is also dependent on the transcription factor RUNX1. RUNX1 null mutant embryos undergo primitive yolk sac hematopoiesis normally, but die between E11.0 and E12.0 due to a failure of definitive hematopoiesis.[83] *LacZ* knock-in mice revealed expression of *Runx1* in a subpopulation of cells in the yolk sac endothelium and in the floor of the dorsal aorta, suggesting that RUNX1 supports the differentiation of hemogenic endothelium in vivo, which contributes to definitive hematopoiesis.[84] Consistent with this idea, aortic clusters of hematopoietic cells are absent in RUNX1 null mutants,[85] and in vitro RUNX1 null embryoid bodies produce clonal blast colony-forming cells at reduced numbers.[86]

Conclusions and Future Directions

Experimentation in several embryonic model systems by many different laboratories has revealed that VEGF signaling is critical for the formation of an initial vascular plexus from multipotent mesodermal progenitors. Continued VEGF signaling, fine-tuned via its receptors and coreceptors, then modulates endothelial cell behavior to establish a functional circulatory network. Other signaling pathways function coordinately with VEGF to mediate vascular plexus remodeling, endothelial cell specification and mural cell recruitment. The formation of blood vessels and blood cells occurs concomitantly. The VEGF receptor KDR is widely recognized as a common marker of all cells possessing putative hemangioblast properties.[87] Cells that retain KDR activity have endothelial potential, whereas cells that activate hematopoietic transcription factors such as TAL1 and RUNX1 gain hematopoietic activity. A subset of KDR-expressing cells that are TAL1+ may represent primitive hemangioblasts capable of initiating primitive yolk sac erythropoiesis, while KDR-expressing cells that are both TAL+ and RUNX1+ represent definitive precursors for definitive hematopoiesis in the AGM and yolk sac.[78] This model of successive diversion of subsets of VEGF-responsive, KDR-expressing cells perhaps best explains our current understanding of the formation of endothelial, primitive hematopoietic and definitive hematopoietic precursors in the early embryo. However, continued research is needed to fully understand the complexity of VEGF signaling. Moreover, we need to increase our efforts to elucidate how VEGF signaling cooperates with other signaling pathways during vascular development and hematopoiesis. Insights gained from ongoing work in developmental model systems will likely benefit the optimization of clinical therapies for those prevalent diseases that are associated with disrupted blood cell and blood vessel formation.

References

1. Tam PPL, Behringer RR. Mouse gastrulation: the formation of a mammalian body plan. Mech Dev 1997; 68:3-25.
2. Winnier G, Blessing M, Labosky PA, Hogan BL. Bone morphogenetic protein-4 is required for mesoderm formation and patterning in the mouse. Genes Dev 1995; 9:2105-2116.
3. Saxton TM, Pawson T. Morphogenetic movements at gastrulation require the SH2 tyrosine phosphatase Shp2. Proc Natl Acad Sci U S A 1999; 96:3790-3795.
4. Dyer MA, Farrington SM, Mohn D et al. Indian hedgehog activates hematopoiesis and vasculogenesis and can respecify prospective neurectodermal cell fate in the mouse embryo. Development 2001; 128:1717-1730.
5. Poole TJ, Finkelstein EB, Cox CM. The role of FGF and VEGF in angioblast induction and migration during vascular development. Dev Dyn 2001; 220:1-17.
6. Flamme I, Risau W. Induction of vasculogenesis and hematopoiesis in vitro. Development 1992; 116:435-439.
7. Motoike T, Loughna S, Perens E et al. Universal GFP reporter for the study of vascular development. Genesis 2000; 28:75-81.
8. Damert A, Miquerol L, Gertsenstein M et al. Insufficient VEGF-A activity in yolk sac endoderm compromises haematopoietic and endothelial differentiation. Development 2002; 129:1881-1892.
9. Carmeliet P, Ferreira V, Breier G et al. Abnormal blood vessel development and lethality in embryos lacking a single VEGF allele. Nature 1996; 380:435-440.
10. Shalaby F, Rossant J, Yamaguchi TP et al. Failure of blood-island formation and vasculogenesis in Flk1-deficient mice. Nature 1995; 376:62-67.
11. Drake CJ, LaRue A, Ferrara N, Little CD. VEGF regulates cell behavior during vasculogenesis. Dev Biol 2000; 224:178-188.
12. Pardanaud L, Luton D, Prigent M et al. Two distinct endothelial lineages in ontogeny, one of them related to hemopoiesis. Development 1996; 122:1363-1371.
13. Cleaver O, Tonissen KF, Saha MS, Krieg PA. Neovascularization of the Xenopus embryo. Dev Dyn 1997; 210:66-77.
14. Soker S, Takashima S, Miao HQ et al. Neuropilin-1 is expressed by endothelial and tumor cells as an isoform-specific receptor for vascular endothelial growth factor. Cell 1998; 92:735-745.
15. Chen H, Chedotal A, He Z et al. Neuropilin-2, a novel member of the neuropilin family, is a high affinity receptor for the semaphorins Sema E and SemaIV but not Sema III. Neuron 1997; 19:547-559.

16. Kolodkin AL, Levengood DV, Rowe EG et al. Neuropilin is a semaphorin III receptor. Cell 1997; 90:753-762.
17. Gluzman-Poltorak Z, Cohen T, Herzog Y, Neufeld G. Neuropilin-2 is a receptor for the vascular endothelial growth factor (VEGF) forms VEGF-145 and VEGF-165. J Biol Chem 2000; 275:18040-18045.
18. Kawasaki T, Kitsukawa T, Bekku Y et al. A requirement for neuropilin-1 in embryonic vessel formation. Development 1999; 126:4895-4902.
19. Chen H, Bagri A, Zupicich JA et al. Neuropilin-2 regulates the development of selective cranial and sensory nerves and hippocampal mossy fiber projections. Neuron 2000; 25:43-56.
20. Takashima S, Kitakaze M, Asakura M et al. Targeting of both mouse neuropilin-1 and neuropilin-2 genes severely impairs developmental yolk sac and embryonic angiogenesis. Proc Nat Acad Sci U S A 2002; 99:3657-3662.
21. Wong C, Jin ZG. Protein kinase Calpha-dependent protein kinase D activation modulates ERK signal pathway and endothelial cell proliferation by VEGF. J Biol Chem 2005; Epub ahead of print.
22. Leung DW, Cachianes G, Kuang WJ et al. Vascular endothelial growth factor is a secreted angiogenic mitogen. Science 1989; 246:1306-1309.
23. Guo D, Jia Q, Song HY et al. Vascular endothelial growth factor promotes tyrosine phosphorylation of mediators of signal transduction that contain SH2 domains. Association with endothelial cell proliferation. J Biol Chem 1995; 270:6729-6733.
24. Yoshida A, Anand-Apte B, Zetter BR. Differential endothelial migration and proliferation to basic fibroblast growth factor and vascular endothelial growth factor. Growth Factors 1996; 13:57-64.
25. Miquerol L, Gertsenstein M, Harpal K et al. Multiple developmental roles of VEGF suggested by a LacZ-tagged allele. Dev Biol 1999; 212:307-322.
26. Cleaver O, Krieg PA. VEGF mediates angioblast migration during development of the dorsal aorta in Xenopus. Development 1998; 125:3905-3914.
27. Lindner V, Majack RA, Reidy MA. Basic fibroblast growth factor stimulates endothelial regrowth and proliferation in denuded arteries. J Clin Invest 1990; 85:2004-2008.
28. Seghezzi G, Patel S, Ren CJ et al. Fibroblast growth factor-2 (FGF-2) induces vascular endothelial growth factor (VEGF) expression in the endothelial cells of forming capillaries: an autocrine mechanism contributing to angiogenesis. J Biol Chem 1998; 141:1659-1673.
29. Goto F, Goto K, Weindel K, Folkman J. Synergistic effects of vascular endothelial growth factor and basic fibroblast growth factor on the proliferation and cord formation of bovine capillary endothelial cells within collagen gels. Lab Invest 1993; 69:508-517.
30. Cai J, Jiang WG, Ahmed A, Boulton M. Vascular endothelial growth factor-induced endothelial cell proliferation is regulated by interaction between VEGFR-2, SH-PTP1 and eNOS. Microvasc Res 2005; 71:20-31.
31. Zeng H, Dvorak HF, Mukhopadhyay D. Vascular permeability factor (VPF)/vascular endothelial growth factor (VEGF) receptor-1 down-modulates VPF/VEGF receptor-2-mediated endothelial cell proliferation, but not migration, through phosphatidylinositol 3-kinase-dependent pathways. J Biol Chem 2001; 276(29):26969-26979.
32. Fong GH, Rossant J, Gertsenstein M, Breitman ML. Role of the Flt-1 receptor tyrosine kinase in regulating the assembly of vascular endothelium. Nature 1995; 376:66-71.
33. Fong GH, Zhang L, Bryce DM, Peng J. Increased hemangioblast commitment, not vascular disorganization, is the primary defect in flt-1 knock-out mice. Development 1999; 126:3015-3025.
34. Lai L, Bohnsack BL, Niederreither K, Hirschi KK. Retinoic acid regulates endothelial cell proliferation during vasculogenesis. Development 2003; 130:6465-6474.
35. Bohnsack BL, Lai L, Dolle P, Hirschi KK. Signaling hierarchy downstream of retinoic acid that independently regulates vascular remodeling and endothelial cell proliferation. Genes Dev 2004; 18:1345-1358.
36. Wasserman SM, Mehraban F, Komuves LG et al. Gene expression profile of human endothelial cells exposed to sustained fluid shear stress. Physiol Genomics 2002; 12:13-23.
37. Brooks AR, Lelkes PI, Rubanyi GM. Gene expression profiling of human aortic endothelial cells exposed to disturbed flow and steady laminar flow. Physiol Genomics 2002; 9:27-41.
38. Chen BPC, Li Y-S, Zhao Y et al. DNA microarray analysis of gene expression in endothelial cells in response to 24-h shear stress. Physiol Genomics 2001; 7:55-63.
39. McCormick SM, Eskin SG, McIntire LV et al. DNA microarray reveals changes in gene expression of shear stressed human umbilical vein endothelial cells. Proc Natl Acad Sci U S A 2001; 98:8955-8960.
40. Garcia-Cardena G, Comander J, Anderson KR et al. Biomechanical activation of vascular endothelium as a determinant of its functional phenotype. Proc Natl Acad Sci U S A 2001; 98:4478-4485.

41. Urbich C, Stein M, Reisinger K et al. Fluid shear stress-induced transcriptional activation of the vascular endothelial growth factor receptor-2 gene requires Sp1-dependent DNA binding. FEBS Lett 2003; 535:87-93.
42. Chen KD, Li YS, Kim M et al. Mechanotransduction in response to shear stress. J Biol Chem 1999; 274:18393-18400.
43. Jalali S, Li YS, Sotoudeh M et al. Shear stress activates p60-src-Ras-MAPK signaling pathways in vascular endothelial cells. Arterioscler Thromb Vasc Biol 1998; 18:227-234.
44. Lee HJ, Koh GY. Shear stress activates Tie2 receptor tyrosine kinase in human endothelial cells. Biochem Biophys Res Comm 2003; 304:399-404.
45. Jin ZG, Ueba H, Tanimoto T et al. Ligand-independent activation of vascular endothelial growth factor receptor 2 by fluid shear stress regulates activation of endothelial nitric oxide synthase. Circ Res 2003; 93:354-363.
46. Dimmeler S, Fleming I, Fisslthaler B et al. Activation of nitric oxide synthase in endothelial cells by Akt-dependent phosphorylation. Nature 1999; 399:601-606.
47. Dimmeler S, Assmus B, Hermann C et al. Fluid shear stress stimulates phosphorylation of Akt in human endothelial cells. Circ Res 1998; 83:334-341.
48. Mattiussi S, Turrini P, Testolin L et al. p21$^{Wafl/Cip1/Sdi1}$ mediates shear stress-dependent antiapoptotic function. Cardiovasc Res 2004; 61:693-704.
49. Akimoto S, Mitsumata M, Sasaguri T, Yoshida Y. Laminar shear stress inhibits vascular endothelial cell proliferation by inducing cyclin-dependent kinase inhibitor p21$^{Sdi1/Cip1/Wafl}$. Circ Res 2000; 86:185-190.
50. Silver L, Palis J. Initiation of murine embryonic erythropoiesis: a spatial analysis. Blood 1997; 89:1154-1164.
51. Palis J, Yoder MC. Yolk-sac hematopoiesis: the first blood cells of mouse and man. Exp Hematol 2001; 29:927-936.
52. Dzierzak E, Medvinsky A, de Bruijn M. Qualitative and quantitative aspects of hematopoietic cell development in the mammalian embryo. Immunol Today 1998; 19:228-236.
53. Godin I, Cumano A. The hare and the tortoise: an embryonic haematopoietic race. Nature Rev 2002; 2:593-604.
54. Garcia-Porrero JA, Godin IE, Dieterlen-Lievre F. Potential intraembryonic hemogenic sites at pre-liver stages in the mouse. Anat Embryol 1995; 192:425-435.
55. Nishikawa S-I, Nishikawa S, Kawamoto H. In vitro generation of lymphohematopoietic cells from endothelial cells purified from murine embryos. Immunity 1998; 8:761-769.
56. Hirai H, Ogawa M, Suzuki N et al. Hemogenic and nonhemogenic endothelium can be distinguished by the activity of fetal liver kinase (Flk)-1 promoter/enhancer during mouse embryogenesis. Blood 2003; 101:886-893.
57. Choi K, Kennedy M, Kazarov A et al. A common precursor for hematopoietic and endothelial cells. Development 1998; 125:725-732.
58. Faloon P, Arentson E, Kazarov A et al. Basic fibroblast growth factor positively regulates hematopoietic development. Development 2000; 127:1931-1941.
59. Chung YS, Zhang WJ, Arentson E et al. Lineage analysis of the hemangioblast as defined by FLK1 and SCL expression. Development 2002; 129:5511-5520.
60. Yamaguchi TP, Dumont DJ, Conlon RA et al. Flk-1, a Flt-related receptor tyrosine kinase is an early marker for endothelial cell precursors. Development 1993; 118:489-498.
61. Millauer B, Wizigmann-Voos S, Schnurch H et al. High affinity VEGF binding and developmental expression suggest Flk-1 as a major regulator of vasculogenesis and angiogenesis. Cell 1993; 72:835-846.
62. Kabrun N, Buhring HJ, Choi K et al. Flk-1 expression defines a population of early embryonic hematopoietic precursors. Development 1997; 124:2039-2048.
63. Kallianpur AR, Jordan JE, Brandt SJ. The SCL/Tal-1 gene is expressed in progenitors of both the hematopoietic and vascular systems during embryogenesis. Blood 1994; 83:1200-1208.
64. Breier G, Breviario F, Caveda L et al. Molecular cloning and expression of murine vascular endothelial-cadherin in early stage development of the cardiovascular system. Blood 1996; 87:630-641.
65. Orkin SH. GATA-binding transcription factors in hematopoietic cells. Blood 1992; 80:575-581.
66. Huber TL, Kouskoff V, Fehling HJ et al. Haemangioblast commitment is initiated in the primitive streak of the mouse embryo. Nature 2004; 432:625-630.
67. Nadin BM, Goodell MA, Hirschi KK. Phenotype and hematopoietic potential of side population cells throughout embryonic development. Blood 2003; 102(7):2436-2443.
68. Wang L, Li L, Shojaei F et al. Endothelial and hematopoietic cell fate of human embryonic stem cells originates from primitive endothelium with hemangioblastic properties. Immunity 2004; 21:31-41.

69. Dumont DJ, Fong GH, Puri MC et al. Vascularization of the mouse embryo: a study of Flk-1, Tek, Tie and vascular endothelial growth factor expression during development. Dev Dyn 1995; 203:80-92.

70. Shalaby F, Ho J, Stanford WL et al. A requirement for Flk1 in primitive and definitive hematopoiesis and vasculogenesis. Cell 1997; 89:981-990.

71. Martin R, Lahlil R, Damert A et al. SCL interacts with VEGF to suppress apoptosis at the onset of hematopoiesis. Development 2003; 131:693-702.

72. Schuh AC, Faloon P, Hu Q-L et al. In vitro hematopoietic and endothelial potential of flk-1-/- embryonic stem cells and embryos. Proc Natl Acad Sci U S A 1999; 96:2159-2164.

73. Elefanty AG, Begley CG, Hartley L et al. SCL expression in the mouse embryo detected with a targeted lacZ reporter gene demonstrates its localization to hematopoietic, vascular, and neural tissues. Blood 1999; 94:3754-3763.

74. Robertson SM, Kennedy M, Shannon JM, Keller G. A transitional stage in the commitment of mesoderm to hematopoiesis requiring the transcription factor SCL/Tal-1. Development 2000; 127:2447-2459.

75. Visvader JE, Fujiwara Y, Orkin SH. Unsuspected role for the T-cell leukemia protein SCL/Tal-1 in vascular development. Genes Dev 1998; 12(473-479).

76. Shivdasani RA, Mayer EL, Orkin SH. Absence of blood formation in mice lacking the T-cell leukaemia oncoprotein Tal-1/SCL. Nature 1995; 373:432-434.

77. Robb L, Begley CG. The helix-loop-helix gene SCL: implicated in T-cell acute lymphoblastic leukaemia and in normal haematopoietic development. Int J Biochem Cell Biol 1996; 28:609-618.

78. Ema M, Faloon P, Zhang WJ et al. Combinatorial effects of Flk1 and Tal1 on vascular and hematopoietic development in the mouse. Genes Dev 2003; 17:380-393.

79. Fujiwara Y, Browne CP, Cunniff K et al. Arrested development of embryonic red cell precursors in mouse embryos lacking transcription factor GATA-1. Proc Natl Acad Sci U S A 1996; 93:12355-12358.

80. Tsai FY, Keller G, Kuo FC et al. An early haematopoietic defect in mice lacking the transcription factor GATA-2. Nature 1994; 371:221-226.

81. Yamada Y, Pannell R, Forster A, Rabbitts TH. The oncogenic LIM-only transcription factor Lmo2 regulates angiogenesis but not vasculogenesis in mice. Proc Natl Acad Sci U S A 2000; 97(1):320-324.

82. Warren AJ, Colledge WH, Carlton MB et al. The oncogenic cysteine-rich LIM domain protein rbtn2 is essential for erythroid development. Cell 1994; 78:45-57.

83. Okuda T, van Deursen J, Hiebert SW et al. AML1, the target of multiple chromosomal translocations in human leukemia, is essential for normal fetal liver hematopoiesis. Cell 1996; 84:321-330.

84. North TE, de Bruijn MF, Stacy T et al. Runx1 expression marks long-term repopulating hematopoietic stem cells in the midgestation mouse embryo. Immunity 2002; 16(5):661-672.

85. North T, Gu TL, Stacy T et al. Cbfa2 is required for the formation of intra-aortic hematopoietic clusters. Development 1999; 126:2563-2575.

86. Lacaud G, Gore L, Kennedy M et al. Runx1 is essential for hematopoietic commitment at the hemangioblast stage of development in vitro. Blood 2002; 100:458-466.

87. Ema M, Rossant J. Cell fate decisions in early blood vessel formation. Trends Cardiovasc Med 2003; 13(6):254-259.

CHAPTER 5

Role of VEGF in Organogenesis

Jody J. Haigh*

Abstract

The cardiovascular system, consisting of the heart, blood vessels and hematopoietic cells, is the first organ system to develop in vertebrates and is essential for providing oxygen and nutrients to the embryo and adult organs. Work done predominantly using the mouse and zebrafish as model systems has demonstrated that Vascular Endothelial Growth Factor (VEGF, also known as VEGFA) and its receptors KDR (FLK1/VEGFR2), FLT1 (VEGFR1), NRP1 and NRP2 play essential roles in many different aspects of cardiovascular development, including endothelial cell differentiation, migration and survival as well as heart formation and hematopoiesis. This review will summarize the approaches taken and conclusions reached in dissecting the role of VEGF signalling in vivo during the development of the early cardiovasculature and other organ systems. The VEGF-mediated assembly of a functional vasculature is also a prerequisite for the proper formation of other organs and for tissue homeostasis, because blood vessels deliver oxygen and nutrients and vascular endothelium provides inductive signals to other tissues. Particular emphasis will therefore be placed in this review on the cellular interactions between vascular endothelium and developing organ systems, in addition to a discussion of the role of VEGF in modulating the behavior of nonendothelial cell populations.

Key Messages
- VEGF plays a central role in the development of the cardiovasculature.
- VEGF supports organogenesis indirectly by promoting vascular development.
- VEGF supports organogenesis directly by acting on various non-endothelial cell types.

Introduction

VEGF is initially expressed at high levels in the yolk sac and in embryonic sites of vessel formation to support the assembly of a cardiovascular system. However, in the embryo, VEGF continues to be expressed after the onset of blood vessel formation, consistent with the idea that it is instructive not only for cardiovascular development, but also for the development of other organ systems. In the adult, VEGF expression becomes restricted to specialized cell types in organs containing fenestrated endothelium, for example the kidney and pituitary.[1] In addition, VEGF is up-regulated to mediate physiological angiogenesis during menstruation, ovulation and in wound healing.[2] The expression of VEGF in adults is also induced by environmental stress caused by hypoxia, anemia, myocardial ischemia, and tumor progression to initiate neovascularisation.[2] Hypoxia up-regulates transcription of the gene encoding VEGF (*Vegfa*) by activating the hypoxia-inducible factors HIF1A and HIF2A (formerly known as HIF1 alpha

*Jody J. Haigh—Vascular Cell Biology Unit, Department for Molecular Biomedical Research (DMBR), Flanders Interuniversity Institute for Biotechnology (VIB)/Ghent University, Technologiepark 927, B9052 Ghent (Zwijnaarde), Belgium. Email: jody.haigh@dmbr.ugent.be

VEGF in Development, edited by Christiana Ruhrberg. ©2008 Landes Bioscience and Springer Science+Business Media.

and HIF2 alpha; see Chapter by 3 M. Fruttiger).[3,4] Several major growth factors up-regulate VEGF, including epidermal growth factor (EGF), insulin-like growth factor (IGF1), fibroblast growth factor 2 (FGF2; also known as basic FGF), fibroblast growth factor 7 (FGF7; formerly known as keratinocyte growth factor, KGF), platelet-derived growth factor (PDGF) and the tumour growth factors TGFA and TGFB (formerly known as TGF alpha and beta).[5] In addition, inflammatory cytokines and hormones have been shown to induce VEGF expression.[6,7]

The mouse VEGF gene (*Vegfa*) gives rise by alternative mRNA splicing and proteolytic processing to three major isoforms termed VEGF120, VEGF164 and VEGF188 (see Chapter 1 by Y.-S. Ng).[8] The corresponding human VEGF isoforms are one amino acid residue larger and therefore termed VEGF121, VEGF165 and VEGF189. In addition, a number of less abundant human isoforms have been described, including VEGF145 VEGF183, VEGF206. The VEGF isoforms differ not only in their molecular mass, but also in their solubility and receptor binding characteristics. VEGG188/VEGF189 and VEGF206 contain exons 6 and 7, which encode heparin-binding domains and mediate binding to heparan sulphate proteoglycan (HSPG) proteins in the extracellular matrix (ECM), and on the cell surface and ECM. VEGF120/VEGF121 lacks the domains encoded by exons 6 and 7 and is therefore the most diffusible isoform. VEGF164/VEGF165 contains the domain encoded by exon 7, but not exon 6, and accordingly has intermediate properties; 50-70% of this isoform remains associated with the cell surface and ECM. Due to the fact that these isoforms differ in their potential to bind to HSPGs, it is believed that they are differentially distributed in the environment of VEGF-secreting cells. The highest levels of VEGF188 are detected in organs that are vascularized initially by vasculogenesis (e.g., lung, heart and liver), while organs vascularized primarily by angiogenesis, including brain, eye, muscle and kidney have higher levels of VEGF164 and VEGF120.[9] The functional roles of these isoforms in organogenesis are described in more detail below.

Approaches to Study VEGF during Organogenesis

Exon 3 is contained in all known VEGF isoforms. The targeted deletion of a single *Vegfa* allele by neomycin gene insertion into this exon leads to embryonic lethality due to abnormal blood vessel and heart development between E11 and E12, demonstrating haploinsufficiency.[10,11] Using high G418 selection of the *Vegfa*+/- neomycin-resistant embryonic stem (ES) cells, *Vegfa*-/- ES cells could be produced, and these were used to make *Vegfa*-/- embryos by tetraploid embryo-ES cell complementation approaches. The resultant *Vegfa*-/- embryos were even more severely compromised than *Vegfa*+/- embryos and died between E9.5 and E10 owing to severe defects in vasculogenesis, angiogenesis and heart development, which was accompanied by tissue necrosis/apoptosis.[10]

Given the relatively early mid-gestational lethality associated with targeted VEGF gene inactivation, alternative strategies have been employed to examine the role of VEGF during organogenesis and in the adult: VEGF function has been ablated at the protein level with sequestering antibodies, small molecule inhibitors that interfere with receptor tyrosine kinase signaling, or a soluble truncated chimeric VEGF receptor termed mFLT1(1-3)-IgG, which consists of the FLT1 extracellular domain fused to an IgG-Fc domain and sequesters VEGF protein with high affinity. Ubiquitous inducible ablation of the conditional VEGF allele or administration of mFLT1(1-3)-IgG causes lethality and severe multi-organ abnormalities in neonatal mice.[12] The defects in the lung, heart and kidney compromised the health status of these two types of mice severely, and the specific functions of VEGF in other organs were therefore difficult to deduce.

To achieve a partial loss of VEGF function, different conditional *Vegfa* alleles containing paired *LoxP* sites have been created, which can be targeted with CRE recombinase to impair VEGF expression in a spatiotemporally defined manner (*Cre/LoxP* technology). Two different *Cre/LoxP* approaches have been used. Firstly, mice in which exon 3 of VEGF is flanked by *LoxP* sites have been generated.[12] In these mice, CRE-mediated excision deletes exon 3, which is present in all VEGF isoforms, and this results in loss of VEGF production.[10,11] This conditional

Vegfa allele can be used to delete VEGF in a cell/tissue specific manner when mated to mouse strains expressing CRE recombinase under an appropriate promoter. An alternative approach was taken to understand the individual roles of the main VEGF isoforms (see Chapter 1 by Y.S. Ng). To create mice expressing only VEGF120, exons 6 and 7 of the *Vegfa* gene were flanked by *LoxP* sites and deleted in ES cells by CRE-mediated excision.[13] These targeted ES cells were then used to make mice expressing VEGF120, but not VEGF164 or VEGF188. To create complementary mice expressing VEGF164 or VEGF188 only, a knock-in approach was used: cDNA sequences corresponding to exons 4,5,7,8 (in the case of VEGF164) or 4-8 (in the case of VEGF188) were inserted into the genomic *Vegfa* locus to remove intervening intron sequences and therefore abolish alternative splicing.[14] Characterization of the phenotypical alterations present in the VEGF isoform-specific mice and the cell/tissue-specific VEGF mutants has contributed greatly to our understanding of VEGF function in organogenesis and in adult life.

Mice that express only the VEGF120 isoform (*Vegfa120/120* mice) make normal overall VEGF levels, which permit the formation of sufficient numbers of endothelial cells and therefore vasculogenesis. Accordingly, these mice survive to birth. However, they do show defects in many organ systems, including the heart,[13,15] retina,[14,16,17] bone,[18,19] kidney[20] and lung.[9] Most *Vegfa120/120* mice die neonatally due to impaired cardiac function, with only 0.5% of animals living to postnatal day 12.[13] An important consideration in interpreting the defects of *Vegfa120/120* mice is that they do not only lack VEGF164 and VEGF188, but also make an increased amount of VEGF120 to compensate for the loss of the other two major splice forms. It has therefore been suggested that the relative balance of the different VEGF isoforms, rather than their absolute levels, is critical for angiogenic vessel patterning.[16] About half of the *Vegfa188/188* mice die prenatally due to severe defects in aortic arch remodeling, and the surviving half show defects in artery development in the eye and defects in epiphyseal bone vascularization.[14,15,21] In contrast to *Vegfa120/120* and *Vegfa188/188* mice, mice that only express VEGF164 (*Vegfa164/164*) or compound *Vegfa120/188* mice develop normally with no discernible phenotypes.[14,16]

The vascular phenotypes of VEGF isoform mutants are likely related to the differential localization of VEGF isoforms in the extracellular space, which provides a control point for regulating vascular branching morphogenesis.[16] Microvessels from mouse embryos expressing only the VEGF120 isoform have an abnormally large diameter and exhibit a decrease in branch formation, which is linked to a delocalization of secreted VEGF protein in the extracellular space. In contrast, mice expressing only the heparin-binding VEGF188 isoform exhibit ectopic vessel branches, which appear long and thin. Intriguingly, the vessel branching and morphogenesis defects of *Vegfa120/120* and *Vegfa188/188* mice are rescued in compound mutant offspring of these mice that express both the VEGF120 and VEGF188 isoforms, even though they lack VEGF164 (*Vegfa120/188* mice). It therefore appears that the growing vasculature migrates along positional guidance cues provided by angiogenic VEGF gradients, which are normally composed of soluble and matrix-binding VEGF isoforms. This VEGF gradient can either be established by the coexpression of VEGF120 and VEGF188 or by VEGF164 alone, as VEGF164 is able to diffuse from the secreting cell, but also binds to and cooperatively signals with extracellular matrix molecules such as heparin sulfate proteoglycans. At the cellular level, VEGF gradients promote the formation of endothelial "tip-cells" at the front of growing vessels, which extend long filopodia presumably to seek out highest VEGF concentrations.[16,17]

Differential VEGF isoform expression does not only control developmental angiogenesis, but is likely to contribute also to physiological and pathological vessel growth in human adults. As an example, the different VEGF isoforms are differentially expressed in humans after submaximal exercise.[22] Moreover, a novel splice variant of VEGF165 termed VEGF165b may provide a natural inhibitor of VEGF165-induced angiogenesis, but is down-regulated in renal cell carcinoma, possibly supporting the switch from an anti-angiogenic to a pro-angiogenic phenotype during tumour development.[23] It has recently been demonstrated that VEGF165, but not VEGF121 specifically interacts with β-amyloid plaques in Alzheimer's disease and may prevent β-amyloid- induced formation of neurotoxic reactive oxygen species.[24]

In order to increase our understanding of the individual roles of the three main VEGF isoforms, we have conditionally targeted each of these isoforms to the ubiquitously expressed *Rosa26* locus, creating mice in which any one of the three isoforms can be selectively overexpressed in a cell type-specific fashion (Haigh et al, unpublished). Transgene expression is under the transcriptional control of the endogenous *Rosa26* promoter, therefore *Cre/LoxP* mediated excision of the *LoxP* flanked stop cassette induces expression of each of the VEGF isoforms at the same level, and the activity of the individual isoforms can therefore be directly compared. We are presently dissecting the individual roles of each of these isoforms using several tissue-specific *Cre* lines to uncover novel roles for the VEGF isoforms in organogenesis and disease.

VEGF Signalling in the Developing Cardiovasculature

The phenotypes of the VEGF and VEGF receptor knockouts all demonstrate that VEGF signalling is critically important to promote the differentiation, proliferation, migration and survival of endothelial cells both during development in the adult. In addition, VEGF signalling plays several distinct roles during heart development. In the following paragraphs, we will first describe the cardiovascular phenotypes of the VEGF receptor mutants and then discuss in more detail the role of VEGF signalling in the nonendothelial cell types of the cardiovasculature and the hematopoetic system (Fig. 1); see also Chapter 3 by M. Fruttiger; Chapter 4 by L.C. Goldie, M.K. Nix and K.K. Hirschi; and Chapter 6 by H. Gerhardt).

Formation of the Cardiovasculature

Formation of Blood Vessels

Migrating mesodermal progenitors that are commonly referred to as hemangioblasts arise soon after gastrulation and the formation of the primitive streak.[25] They establish the blood islands of the extra-embryonic yolk sac, which consist of elongated peripheral endothelial precursor cells (angioblasts) that surround hematopoietic progenitors (hemoblasts). The angioblasts coalesce into primitive vascular channels in a process termed vasculogenesis.[25] Subsequently, the vascular endothelium within these early vessels proliferates and migrates to give rise to new vessel sprouts in a process termed angiogenesis. Vasculogenesis and angiogenesis are repeated in the embryo proper. Nascent vessels subsequently remodel and maturate, which involves the recruitment of pericytes or smooth muscle cells and the development of supporting basement membranes.[26]

Formation of Hematopoetic Stem Cells

In mammals, yolk sac blood islands produce primitive erythrocytes and macrophages as well as progenitors that move on to populate an intra-embryonic region giving rise to definitive hematopoietic stem cells (HSCs); this region is known as the para-aortic splanchopleura (PAS) or aorta-gonads-mesonephros (AGM). The embryonic HSCs populate first the fetal liver and then the bone marrow, which, together with thymus and spleen, produces the hematopoietic cells of the eyrthroid, myeloid and lymphoid cell lineages in the adult.[27]

Formation of the Heart

The development of the heart involves a complex series of finely orchestrated cellular and molecular interactions that commences just after gastrulation.[28] Initially, the primitive cardiac mesoderm ingresses along with other mesodermal and endodermal progenitors through the primitive streak and migrates to the anterior-most region of the embryo. Here, the mesodermal progenitors condense between embryonic day (E) 7.0 and 7.5 to form a crescent shaped epithelium often referred to as the cardiac crescent. The endocardium (future endothelium of the heart) invades the myocardium (the future muscle layer of the heart) and the sub-adjacent endoderm. From around E8.0, the bilateral cardiac progenitors coalesce at the ventral midline and fuse to form a linear heart tube consisting of an inner endocardial tube surrounded by a myocardial epithelium. Between E8.0 and E8.5, this linear heart tube begins a series of looping

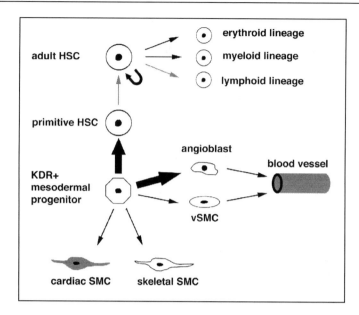

Figure 1. Role of VEGF in endothelial, hematopoetic and muscle differentiation. In the embryo, VEGF signalling promotes differentiation of a KDR+ progenitor cells into the endothelial and hematopoetic lineages. Differentiation decisions skewing progression into these lineages may depend on VEGF signalling acting in concert with transcriptional regulators. Accordingly, VEGF signalling together with TAL1 is thought to promote a hematopoietic/ endothelial cell fate development. VEGF plays an autocrine role in adult hematopoetic stem cells, promoting both survival and differentiation. Pathological overexpression of VEGF may drive HSC differentiation towards the myeloid or erythroid lineages.VEGF also promotes vessel assembly by stimulating the migration of angioblasts and vascular smooth muscle cells, and by supporting vascular smooth muscle cell recruitment into nascent vessels. VEGF signaling may also promote the formation of two other muscle cell types, cardiac and skeletal muscle; in adults, VEGF supports aspects of muscle cell function, such as protection of muscle cells from ischemia, promoting muscle regeneration and regulating cardiac muscle contractility.

morphogenic events that sets the stage for the formation and positioning of the cardiac chambers that begin functioning to pump blood through the remodeling developing vasculature.

Requirements for VEGF Receptors in the Formation of a Cardiovascular Circuit

KDR (FLK1, VEGFR2)

The essential role of VEGF signaling in the initial stages of cardiovascular development is underscored by the fact that KDR receptor null embryos, which contain the beta-galactosidase gene in frame with the ATG in exon 1, die at around E8.5, even earlier than *Vegfa-/-* embryos; death is the result of the failure of mesodermal progenitors to differentiate into vascular endothelium, endocardium and hematopoietic cells.[29] To better understand the role of KDR in vasculogenesis and hematopoiesis, the developmental potential of KDR null ES cells was studied in an ES cell-diploid embryo chimera approach.[30] These studies demonstrated that KDR signaling is essential for the migration of hemangioblasts from the primitive streak region of the embryo to the yolk sac, where they form blood islands. In vitro differentiation assays showed that KDR null ES cells can differentiate into hematopoietic and endothelial cells in vitro, although at a reduced frequency,[31] suggesting that the mesodermal progenitors of hematopoietic and endothelial cells do not require KDR signaling for their formation or differentiation, but rather for their migration and expansion.

FLT1 (VEGFR1)

The role of FLT1 has been more controversial, as it has been debated whether this VEGF receptor acts as a major transducer of VEGF signals or instead acts as a decoy receptor that modulates the amount of VEGF available to KDR. Recent evidence indicates that the conflicting reports may be due, at least in part, to the fact that the functions and signaling properties of FLT1 differ depending on the developmental stage, cellular context, or alternative splicing. FLT1 can be produced as a full-length receptor with an intracellular tyrosine kinase domain or as a truncated receptor that is either membrane bound via its transmembrane domain or secreted. Mice that are homozygous null for FLT1 die between E8.5 and E9.0 as a result of increased and disorganized blood vessels.[32] This increased and aberrant blood vessel development is caused by an increased commitment of mesodermal progenitors to the endothelial/hematopoetic lineage, which has been postulated to result from excessive KDR signaling in the absence of FLT1.[33] Mice that lack the FLT1 tyrosine kinase domain, but express the extracellular and transmembrane domain of FLT1 develop a normal cardiovasculature, supporting the idea that during development FLT1 signaling is not essential and that the extracellular domain of FLT1 serves to modulate the amount of VEGF that binds to KDR.[34] More recently, mice have been made that lack both the tyrosine kinase domain and transmembrane domain of FLT1 and produce mainly the soluble FLT1 protein.[35] In these mice, VEGF is not recruited efficiently to the plasma membrane and KDR receptor activity is reduced, suggesting that the membrane fixation of FLT1 is essential to recruit VEGF for efficient KDR stimulation. Likely because KDR is necessary for the anterior migration of mesodermal prognitors of the cardiovasculature,[36] these mice suffer from embryonic lethality, particularly in a strain with high endogenous KDR levels.[35] However, a proportion of these mice survive embryonic development, which suggests that the endogenous soluble form of FLT1, termed sFLT1, is able to fulfill some functions encoded by the full-length gene. For example, it has been suggested that sFLT1 is critical to shape VEGF gradients in the environment of sprouting blood vessels.[37] In adults, FLT1 is thought to amplify VEGF-signaling through KDR by cooperative binding of VEGF and the other FLT1-specific ligand, placental growth factor (PGF; also known as PlGF). This signalling pathway has been suggested to operate in bone marrow-derived circulating endothelial progenitors, which play important roles in neo-angiogenesis in the adult.[38]

NRP1 and NRP2

The neuropilins NRP1 and NRP2 are transmembrane glycoproteins with adhesive properties that also function as axonal receptors for neuronal guidance cues of the semaphorin family and for specific VEGF isoforms on endothelial and tumor cells; in blood vessels, they are thought to amplify VEGF signaling through KDR.[39,40] Targeted deletion of NRP1 results in embryonic lethality around E12.5 due to defects in aortic arch remodeling, and impaired blood vessel invasion into the neural tube.[41] Compound homozygous null NRP1/NRP2 embryos die between E8.5 and E9.0 due to defective yolk sac vascularisation and are intermediate in severity of cardiovascular defects relative to VEGF and KDR homozygous null embryos.[42] Mice that are either compound homozygous for NRP1 deficiency and heterozygous for NRP2 deficiency or vice versa die around E10.5, demonstrating gene dosage sensitivity as previously demonstrated for VEGF.[42]

Role of VEGF Signalling in Heart Development

A small proportion of *Vegfa120/120* mice survive to term, but they then succumb to cardiac defects that resemble ischemic cardiomyopathy as a result of impaired myocardial angiogenesis.[13] The hearts of the *Vegfa120/120* mice are enlarged and show an irregular heartbeat with dysmorphic, weak contractions. To address the paracrine role of the myocardial VEGF source, mice carrying the conditionally targeted *Vegfa* allele[12] were crossed with the *MLC2v-Cre* transgenic line[43] to specifically delete VEGF in the ventricular myocardium.[44] The observed cardiac phenotypes were similar to the ones seen in *Vegfa120/120* mice, implying that the myocardium provides a critical source of VEGF to support the development of the coronary vasculature.

Earlier in development, VEGF is specifically expressed in the atrioventricular (AV) field of the heart tube, soon after the formation of the endocardial cushions that give rise to the heart septa and valves.[45] In support of the idea that VEGF levels are critical for this aspect of heart formation, 2-3 fold ubiquitous over-expression of VEGF impairs cardiac septation and leads to outflow tract abnormality combined with aberrant trabeculation and myocardial defects.[45] The induction of VEGF in myocardial cells with the tetracycline-inducible system has been used to demonstrate that VEGF can act as a negative regulator of endocardial-to-mesenchymal transformation, which is a critical step in the formation of the endocardial cushions; ischemia-induced VEGF up-regulation has therefore been proposed to play a role in congenital defects of heart septation.[46] *Vegfa120/120* and *Vegfa188/188* mice also display defects in aortic arch remodeling.[15] VEGF signaling through KDR and NRP1 appears to cooperate with plexin D1 and its semaphorin ligand SEMA3C to influence the remodeling of vessels in the developing cardiac outflow tract.[47,48]

Taken together, these studies on the role of VEGF suggest that altering the level of VEGF or the isoform expression profile disrupts vessel formation during heart development at multiple levels. Whilst it was originally thought that KDR is expressed predominantly in the endocardial cells, more recent studies suggest that VEGF/KDR signaling also influences muscle cells at times of cytotoxic stress in the adult, for example when cardiac muscle cells experience hypoxia.[49] Cell fate lineage-marking experiments with the *Rosa26-LacZ* reporter mouse strain and a *Cre* recombinase gene targeted to the *Kdr* locus have demonstrated that KDR+ progenitor cells give rise to cardiac muscle cells and also to skeletal muscle cells (Fig. 1).[50] However, the role of KDR-signaling in modulating the differentiation of KDR+ myogenic progenitor cell populations, in early morphogenic events during heart tube formation and in response to hypoxic stress remains to be genetically determined.

Role of VEGF in Vascular Smooth Muscle Cells

In vitro differentiation of embryonic stem (ES) cells has provided a valuable tool to elucidate the molecular mechanisms involved in cell fate differentiation decisions of KDR+ mesodermal progenitor populations (see Chapter 5 by L.C. Goldie, M.K. Nix and K.K. Hirschi). In particular, ES cell differentiation has been used to identify clonal KDR+ hemangioblast cell populations that can give rise to both endothelial and hematopoietic progenitors.[51] More recently, it has been demonstrated that KDR+ cells derived from ES cells can differentiate into both endothelial and smooth muscle actin-positive cells, which assemble into blood vessels when cultured on an appropriate collagen matrix.[52] The addition of VEGF to these ES cell cultures enhanced the formation of endothelial cells, whereas the addition of PDGFBB enhanced the formation of smooth muscle cells. The ES-cell derived KDR+ cells could also give rise to both vascular endothelium and smooth muscle cells in xenograft models.[52] An ES cell-based study has recently looked at the combinatorial effects of KDR and the transcriptional regulator TAL1 during vascular and hematopoietic development. In this study, KDR+ mesodermal progenitor cells and hemangioblast colony forming cells gave rise to smooth muscle cells, and TAL1 inhibited the formation of smooth muscle cells; moreover, expression of TAL1 in a KDR null background rescued both endothelial and hematopoietic differentiation. This work along with other experiments has lead to a model in which KDR marks a mesodermal precursor cell population that forms KDR+ progeny, but can also give rise to nonKDR expressing cell types, such as smooth muscle, cardiac and skeletal muscle cells[53] (Fig. 1). Mesodermal cells that retain KDR expression have a strong potential to develop into vascular endothelial cells, or may adopt hematopoietic activity in the presence of transcriptional regulators such as TAL1 and RUNX1.[53] The presence of KDR on cultured vascular smooth muscle cells was suggested to stimulate their migration in response to VEGF.[54]

VEGF Signaling in the Hematopoietic System

As described above, VEGF signaling is essential during the initial stages of hematopoietic differentiation in the embryo. To determine the role of VEGF in adult hematopoietic stem cells (HSCs), conditional null VEGF transgenic mice were crossed to mice carrying a *Cre* recombinase transgene under the control of an alpha-interferon inducible promoter (*MX-Cre*).[55] This allowed the selective ablation of VEGF in bone marrow cell isolates and HSCs after administration of α-interferon to the culture medium. In these experiments, VEGF-deficient HSCs and bone marrow mononuclear cells failed to repopulate lethally irradiated hosts, despite the coadministration of a large excess of wild-type supporting cells. These findings demonstrated that a wild-type stromal cell microenvironment is insufficient to rescue engraftment of VEGF-deficient HSC in the bone marrow. VEGF therefore controls HSC survival during hematopoietic repopulation via an internal autocrine-loop mechanism (Fig. 1). VEGF-deficient bone marrow cells also failed to form colonies in vitro, but this defect was rescued by addition of VEGF to the culture medium. A similar rescue effect was observed in vivo, when VEGF-deficient HSCs were transduced with retroviral vectors expressing VEGF, as this restored their ability to repopulate lethally irradiated host mice.

The use of VEGF-selective small molecule receptor antagonists, which can enter the HSC, suggested that both KDR and FLT1 play a role in the VEGF repopulation potential of the HSC.[55] However, in another study it was found that adult mouse hematopoietic stem cells are normally KDR⁻ or KDR^low, and that KDR⁺ cells have no long-term reconstitution capacity.[56] These observations may be reconciled in a model in which KDR is essential for the development of HSCs during early embryonic development, but is redundant with FLT1 for HSCs in adult mouse bone marrow. In agreement with this idea, the enforced expression of PGF in VEGF-deficient bone marrow cells revealed that activation of FLT1 is fully sufficient to rescue HSC survival in vitro and hematopoietic repopulation in vivo.[55] Further support for a model in which FLT1 is required during bone marrow repopulation was provided by experiments using monoclonal antibodies blocking murine FLT1.[57] In order to resolve the controversy surrounding the essential role of the VEGF receptors in HSC function, similar cell specific ablation experiments will need to be performed using conditional null receptor alleles. To determine the requirement for VEGF and its receptors in more mature hematopoietic cell populations, mice carrying conditional VEGF/VEGF-receptor alleles have to be mated to hematopoietic lineage-restricted *Cre* lines.

High levels of systemic VEGF can compromise the adult immune system (Fig. 1). For example, long-term continuous-infusion of recombinant VEGF protein or administration of adenoviral VEGF-expressing vectors to adult mice inhibits dendritic cell development, but increases the production of B cells and immature myeloid cells.[58] Moreover, continuous administration of recombinant VEGF or the ubiquitous induction of VEGF in adult mice at pathological levels mimics the profound thymic atrophy observed in tumor-bearing mice and inhibits the production of T-cells.[58,59] These effects may contribute to the ability of tumors to evade normal immune surveillance. The molecular mechanisms behind these adverse effects in the hematopoietic system remains poorly understood, underscoring the need to unravel the effects of VEGF signalling in the hematopoietic system.

VEGF and the Development of Endoderm-Derived Organs

The ability of vascular endothelium and therefore VEGF signaling to modulate the development of two endoderm-derived organs has been well documented: in two recent hallmark publications, the vascular endothelium was demonstrated to play an essential role in the induction of both liver and pancreas.[60,61]

Liver

In a novel liver primordial explant system, it was demonstrated that there is a failure of liver morphogenesis in KDR-deficient tissues, which lack endothelium, and it was concluded that vasculogenic endothelial cells/nascent blood vessels release critical mediators essential for the earliest stages of hepatogenesis.[60] The effects of VEGF on the liver may not be limited to liver endothelial cells, as there have been reports describing VEGF receptors on liver stellate cells,[62] which play a role in liver regeneration in response to injury. High levels of genetically induced or tumor-produced VEGF can change the architecture of the adult liver, resulting in a liver 'peliosis-like' phenotype that is characterized by enlarged hepatic sinusoids, blood pooling, detached sinusoidal endothelial cells and a total disruption of normal liver architecture.[59,63] This phenotype recapitulates the liver pathologies seen in some cancer patients with a high tumor burden. On the other hand, systemically administered VEGF at moderate levels can act on the sinusoidal vasculature (via FLT1) to enhance secretion of hepatocyte growth factor (HGF), which has a protective effect on hepatocytes under times of cytotoxic stress.[64]

Pancreas

In both frog and transgenic mouse models, blood vessel endothelium induces insulin expression in isolated endoderm; moreover, removal of the dorsal aorta prevents insulin expression in frogs, whilst expression of VEGF under the control of the *Pdx1* promoter causes ectopic vascularization in the posterior foregut, which was accompanied by ectopic insulin expression and islet hyperplasia.[61] These studies underscore the fact that blood vessels are not only essential for metabolic sustenance, but also provide inductive signals essential for organogenesis. The nature of these vessel-derived factors remains to be determined. Recently, VEGF has been conditionally deleted in the endocrine pancreas using a *Pdx1 Cre* line and in β-cells using the RIP *Cre* mouse line.[65,66] Neither approach overtly affected pancreatic islet formation, but affected the fine-tuning of blood glucose regulation.[66] However, the β-cell specific ablation of VEGF in the RIP-Tag2 model of tumorigenesis demonstrated the essential role of VEGF in the angiogenic switch necessary for neoplastic progression.[65]

Lung

That VEGF is an important growth factor during lung maturation was demonstrated by several different mouse mutants, including *Vegfa120/120* mice,[9] mice lacking the VEGF transcriptional regulator HIF2A,[67] and mice carrying a mutation in the hypoxia response element within the 5' UTR of the *Vegfa* gene.[68] The upregulation of the VEGF188 isoform is temporally and spatially associated with the maturation of alveolar epithelium in the lung. The phenotype of mice lacking VEGF188 (*Vegfa120/120* mice) suggested that this isoform is produced by the pulmonary epithelium to mediate the assembly and/or stabilization of the highly organized vessel network, which in turn is critical for alveolar development. Specifically, the lungs of *Vegfa120/120* mice are significantly less developed at birth when compared with to their control littermates, as they show a significant reduction in air space and capillaries; moreover, bronchioles and larger vessels are located unusually close to the pleural surface, pulmonary epithelium is less branched and the formation of both primary and secondary septa is reduced. Lungs from *Vegfa120/120* mice at postnatal day 6 showed similar defects, with retarded alveolarization and capillary growth. Administration of VEGF into the amniotic cavity of prematurely born mice can increase surfactant protein production of type 2 pneumocytes, suggesting that VEGF has therapeutic potential to encourage lung maturation in preterm infants.[67] On the other hand, the constitutive expression of VEGF during development increases the growth of pulmonary vessels, disrupts branching morphogenesis in the lung and inhibits of type I pneumocyte differentiation.[69] Inducible lung-specific VEGF expression in the neonate up till six weeks of age using a tetracycline inducible approach caused pulmonary hemorrhage, hemosiderosis, alveolar remodeling changes, and inflammation.[70] Using a similar VEGF inducible system in the lung with slightly later times of VEGF induction demonstrated that VEGF-induced lung inflammation resembles the lung pathology of patients with asthma, raising the possibility that VEGF inhibition in the lung may be of therapeutic benefit in the treatment of asthma or other helper T cell (TH-2) mediated inflammatory disorders elsewhere in the body.[71]

VEGF in Mesoderm-Derived Organs

Kidney

The main functional unit of the kidney is the glomerular filtration barrier consisting of visceral epithelial cells termed podocytes, fenestrated endothelial cells and an intervening glomerular basement membrane. This filtration barrier is responsible for the purification of the blood; it allows water and small solutes to pass freely into the urinary space, but ensures retention of critical blood proteins such as albumin and blood clotting factors. In the developing glomerulus, all three main VEGF isoforms are highly expressed in presumptive and mature podocytes.[72] Gene targeting in the mouse has demonstrated that VEGF is required for the development and maintenance of the glomerular filtration barrier. Podocyte-specific deletion of one conditional *Vegfa* allele develop symptoms that resemble glomerular endotheliosis, with excessive protein content in the urine.[73] Podocyte-specific deletion of both conditional *Vegfa* alleles leads to congenital nephropathy and perinatal lethality.[73] *Vegfa120/120* mice display smaller glomeruli and have fewer capillary loops.[20] Conversely, podocyte-specific over-expression of VEGF164 causes a phenotype that resembles HIV-associated collapsing glomerulopathy.[73] For a more detailed review of VEGF and its receptors in glomerular development and pathologies see the review article by Eremina and Quaggin.[74]

Skeletal Muscle

As mentioned above, cell fate lineage-marking experiments have demonstrated that KDR+ progenitor cells can give rise to skeletal muscle cells during development (Fig. 1).[50] In the adult, VEGF expression is markedly enhanced in regenerating muscle fibres,[75] and it is thought that VEGF may directly act on skeletal muscle cells.[76] KDR and FLT1 are both expressed by a myoblast cell line termed C2C12. Treatment of C2C12 cells with VEGF results in a significant induction of myoglobin mRNA, and VEGF-induced myoglobin mRNA expression is completely abolished when VEGF treatment was combined with administration of a VEGF-receptor tyrosine kinase inhibitor. In addition, ischemic muscles expressing exogenously administered VEGF vectors become deeply red in color, due predominately to up-regulation of myoglobin expression in the skeletal muscle rather than increased vascularisation or vascular leakage.[75] Given these intriguing findings, VEGF signaling may have a broader role in muscle cell biology than was originally assumed (Fig. 1). Given the fact that VEGF therapy is already being used in several clinical trials for treating hindlimb and cardiac ischemia with mixed degrees of success, more work needs to be done using in vivo models to elucidate the biological role of VEGF in muscle cell development and in pathological conditions.

Bone

During endochondral bone formation, the cartilage anlagen develop as an avascular tissue until around E14.5 in the mouse, when proliferating chondrocytes begin to hypertrophy and secrete pro-angiogenic factors such as VEGF (see Chapter 7 by C. Maes and G. Carmeliet).[77,78] The up-regulation of VEGF correlates with the time of angiogenic invasion into the cartilage anlagen and sets in motion a complex developmental process, in which the cartilage is remodeled into trabecular bone. Inhibition of VEGF by administration of soluble chimeric VEGF receptor protein to 24-day-old mice was found to inhibit blood vessel invasion into the hypertrophic zone of long bone growth plates, impaired trabecular bone formation and caused an expansion of the hypertrophic zone.[77] Initial attempts to conditionally delete VEGF in the developing cartilage using a collagen-2 promoter-driven *Cre* gene resulted in similar phenotypes in heterozygous null mice; however the ectopic expression of CRE in the myocardium in this particular transgenic line compromised VEGF expression during heart development and therefore caused embryonic lethality in homozygously targeted mice[78] (see above). More recently, a different *Cre* line, also driven by the collagen 2 promoter, was used to ablate the conditional VEGF allele specifically and selectively in cartilage; the resultant VEGF mutants lacked angiogenic invasion into the developing cartilage, and, in addition, showed extensive chondrocyte death.[79] This observation raised the possibility that VEGF promotes the survival and differentiation of chondrocytes firstly by attracting a blood supply, and secondly in a cell autonomous mechanism that operates independently of blood vessels. Analysis of *Vegfa120/ 120* mice indicated an essential role of the VEGF164 and VEGF188 isoforms in the initial invasion and remodeling of growth plate during bone formation.[18,19] *Vegfa188/188* mice, on the other hand, display knee joint dysplasia and dwarfism due to disrupted development of the growth plates and secondary ossification centers. At the cellular level, this phenotype is in part caused by impaired vascularization of the epiphysis, which increases hypoxia and therefore causes massive chondrocyte death in the interior of the epiphyseal cartilage. The VEGF188 isoform alone is not only inefficient in promoting epiphyseal vascularisation, but is also insufficient to regulate chondrocyte proliferation and survival responses to hypoxia.[21] Taken together, these studies suggest that VEGF regulates the differentiation of hypertrophic chondrocytes, osteoblasts and osteoclasts. However, the VEGF receptors that regulate bone cell responses to VEGF remain to be identified.

VEGF in Ectoderm-Derived Organs (Skin and Nervous System)

Skin

The main source of VEGF production in the skin is the epidermis, consisting of keratinocytes.[80] In order to determine the role of keratinocyte-derived VEGF in skin development as well as in adult skin function and pathologies, the conditional *Vegfa* allele has been inactivated in epidermal keratinocytes with the *K5 Cre* mouse line, in which CRE recombinase expression is driven by the keratin 5 promoter.[12,81] These skin-specific VEGF mutant mice formed a normal skin capillary system and displayed normal skin and hair development, demonstrating that keratinocyte-derived VEGF is not essential for skin development. However, healing of full-thickness skin wounds in adult animals lacking epidermal VEGF expression was delayed, and these animals showed a marked resistance to skin carcinoma formation induced by chemical or genetic means.[81] Vice versa, over-expression of VEGF in epidermal keratinocytes under the control of the keratin 5 regulatory sequences increased susceptibility to chemically induced carcinogenesis.[82] Other model systems relying on the transgenic delivery of VEGF to the skin have resulted in an inflammatory skin condition with many of the cellular and molecular features of psoriasis.[83] These studies have underscored the contribution of VEGF signalling to several skin pathologies.

Nervous System

There are several intriguing similarities between the nervous system and the vasculature. For example, both organ systems form dense networks throughout the entire body and, in both cases, initially primitive networks undergo remodeling and maturation processes to refine their hierarchical organization and integration with other tissues; moreover, both organ systems use signalling ligands and receptors from shared protein families.[84,85] Of particular interest is the presence of NRP1 on both neurons and endothelial cells, as it plays a pivotal role in neuronal guidance by acting as a semaphorin receptor and also acts as a cofactor for VEGF164-mediated KDR signaling. More recently, the role of semaphorins in the vasculature and of VEGF in the nervous system has become the focus of much research activity. For example, the administration of VEGF to the developing CNS improved intraneural angiogenesis, which in turn promoted neurogenesis, and the application of VEGF to experimental models of nerve injury enhanced nerve regeneration after axotomy.[86,87] Several in vitro experiments demonstrated more direct effects of VEGF in several aspects of neuronal biology, consistent with a model of a direct autocrine effect of VEGF on neurons and glial cells. For example, VEGF was shown to rescue neuronal cultures from ischemic and glutamate induced neurotoxicity and to act like a neurotrophic factor in stimulating axonal outgrowth and glial growth.[88,89]

VEGF/KDR signalling has been targeted in neural progenitor cells and their progeny with a nestin promoter-driven *Cre* recombinase transgene to inactivate a conditional null *Vegfa* allele[12] in combination with a hypomorphic *Vegfa* allele,[90] as well as inactivating a conditional null *Kdr* allele.[91] Mice with intermediate levels of VEGF activity showed decreased blood vessel branching and a reduced vascular density in the cortex and retina, and this impaired the structural organization of the cortex and retinal thickening.[91] Severe reductions in VEGF levels decreased vascularity severely and resulted in hypoxia, which led to the degeneration of the cerebral cortex and caused neonatal lethality.[91,92] In contrast, the deletion of KDR in neurons did not obviously impair neuronal development, suggesting that the major developmental role of VEGF in the nervous system is that of providing a paracrine factor for the vasculature. However, VEGF164 does control specific aspects of neuronal patterning, as it is essential for the migration of facial branchiomotor neuron cell bodies.[93] Interestingly, initial experiments suggest that NRP1, rather than a NRP1/KDR complex, controls the VEGF response in this neuronal migration pathway (Q. Schwarz and C. Ruhrberg, personal communication).

Despite the lack of an obvious neuronal phenotype of mice specifically lacking KDR-expression in the nervous system, we have recently isolated definitive neuronal stem cells (dNSCs) from embryonic and adult brains of these mice and cultured these cells in vitro to produce neurospheres. Interestingly, dNSCs isolated from the KDR null mutant brains showed a 50% reduction in the capacity to form neurospheres and a dramatic increase in the number of dNSCs undergoing apoptosis.[94] These results imply that in an environment where growth factors are limited, KDR plays a cell autonomous role in NSC survival that is normally overcome in vivo, i.e. in the brains of mice lacking KDR.[91] Further support for an important role of environmental issues in VEGF signaling and neuronal function has come from studies that have demonstrated that hippocampal sources of VEGF-A may link neurogenesis with enhanced learning and memory potential.[95] In this study, enriched external environments result in increased hippocampal VEGF expression, increased neurogenesis and improved cognitive performance. Exogenously administered retroviral sources of VEGF alone caused similar affects on learning and memory. This is mediated in part by direct signaling through KDR present on rat hippocampal neurons.

From a clinical perspective, altered VEGF levels and vascular development have been associated with neurodegenerative disorders, ischemic cerebral and spinal cord injury, and diabetic and ischemic neuropathy.[96,97] VEGF has also recently been implicated in playing a causal role in the pathogenesis of amyotrophic lateral sclerosis (ALS).[68] In a mouse model of ALS, injections of lentiviral vectors expressing VEGF into various muscles delayed disease onset and disease progression and extended the survival time by 30% in ALS prone mice.[98] VEGF is thought to modulate neuronal development and function indirectly through its paracrine effects on the vasculature, which provides essential neurotrophic support during nervous system development and homeostasis. More recently, it has been demonstrated that transgenic expression of KDR on motor neurons prolonged the survival of mice that are genetically prone to develop ALS.[99] For a more extensive review on the role of VEGF signaling in the nervous system please refer to Chapter 8 by J.M. Rosenstein, J.M. Krum and C. Ruhrberg.

Future Perspectives

The phenotypical consequences of altering the overall levels of VEGF and/or its isoforms have made it very clear that this growth factor plays many diverse and essential roles in normal organogenesis beyond its initial role in the establishment of the cardiovasculature. It is therefore not surprising that aberrant VEGF signaling has been implicated in various human pathological conditions, including the growth and metastasis of tumors, deregulated immune cell surveillance, intraocular neovascular syndromes, enhanced inflammatory responses in arthritis and brain edema, as well as neurological disorders such as cerebral and spinal trauma, ischemic and diabetic neuropathy and amyotrophic lateral sclerosis. A better understanding of the physiological role of VEGF signaling during normal organogenesis will undoubtedly lead to novel insights into the pathenogenesis of these diseases, and it may even lead to novel therapeutic approaches to combat these life threatening and debilitating diseases. The multitude of phenotypes that have been documented as a result of altering VEGF levels in the mouse have already provided a great deal of insight into several human pathologies ranging from cancer progression to amyotrophic lateral sclerosis. Some of the immediate challenges for the future include: (a) to obtain a better understanding of the molecular mechanism of VEGF signaling plays in vascular endothelium and hematopoietic cells (i. e. identification of direct signaling targets, transcriptional modulators and target genes); and (b) to understand the role of VEGF in modulating nonendothelial/nonhematopoietic cell types (i.e., through the analysis of cell/tissue specific knock-outs of VEGF receptors). More long-term challenges include the molecular characterization of signaling pathways that can modulate the activity of VEGF and the development of effective therapeutic approaches to modulate VEGF signaling in human diseases.

References

1. Miquerol L, Gertsenstein M, Harpal K et al. Multiple developmental roles of VEGF suggested by a LacZ-tagged allele. Dev Biol 1999; 212(2):307-322.
2. Ferrara N. The role of VEGF in the regulation of physiological and pathological angiogenesis. Exs 2005; (94):209-231.
3. Ema M, Taya S, Yokotani N et al. A novel bHLH-PAS factor with close sequence similarity to hypoxia-inducible factor 1alpha regulates the VEGF expression and is potentially involved in lung and vascular development. Proc Natl Acad Sci USA 1997; 94(9):4273-4278.
4. Forsythe JA, Jiang BH, Iyer NV et al. Activation of vascular endothelial growth factor gene transcription by hypoxia-inducible factor 1. Mol Cell Biol 1996; 16(9):4604-4613.
5. Pages G, Pouyssegur J. Transcriptional regulation of the Vascular Endothelial Growth Factor gene—a concert of activating factors. Cardiovasc Res 2005; 65(3):564-573.
6. Fujita M, Mason RJ, Cool C et al. Pulmonary hypertension in TNF-alpha-overexpressing mice is associated with decreased VEGF gene expression. J Appl Physiol 2002; 93(6):2162-2170.
7. Harada S, Nagy JA, Sullivan KA et al. Induction of vascular endothelial growth factor expression by prostaglandin E2 and E1 in osteoblasts. J Clin Invest 1994; 93(6):2490-2496.
8. Ruhrberg C. Growing and shaping the vascular tree: Multiple roles for VEGF. Bioessays 2003; 25(11):1052-1060.
9. Ng YS, Rohan R, Sunday ME et al. Differential expression of VEGF isoforms in mouse during development and in the adult. Dev Dyn 2001; 220(2):112-121.
10. Carmeliet P, Ferreira V, Breier G et al. Abnormal blood vessel development and lethality in embryos lacking a single VEGF allele. Nature 1996; 380(6573):435-439.
11. Ferrara N, Carver-Moore K, Chen H et al. Heterozygous embryonic lethality induced by targeted inactivation of the VEGF gene. Nature 1996; 380(6573):439-442.
12. Gerber HP, Hillan KJ, Ryan AM et al. VEGF is required for growth and survival in neonatal mice. Development 1999; 126(6):1149-1159.
13. Carmeliet P, Ng YS, Nuyens D et al. Impaired myocardial angiogenesis and ischemic cardiomyopathy in mice lacking the vascular endothelial growth factor isoforms VEGF164 and VEGF188. Nat Med 1999; 5(5):495-502.
14. Stalmans I, Ng YS, Rohan R et al. Arteriolar and venular patterning in retinas of mice selectively expressing VEGF isoforms. J Clin Invest 2002; 109(3):327-336.
15. Stalmans I, Lambrechts D, De Smet F et al. VEGF: A modifier of the del22q11 (DiGeorge) syndrome? Nat Med 2003; 9(2):173-182.
16. Ruhrberg C, Gerhardt H, Golding M et al. Spatially restricted patterning cues provided by heparin-binding VEGF-A control blood vessel branching morphogenesis. Genes Dev 2002; 16(20):2684-2698.
17. Gerhardt H, Golding M, Fruttiger M et al. VEGF guides angiogenic sprouting utilizing endothelial tip cell filopodia. J Cell Biol 2003; 161(6):1163-1177.
18. Maes C, Carmeliet P, Moermans K et al. Impaired angiogenesis and endochondral bone formation in mice lacking the vascular endothelial growth factor isoforms VEGF164 and VEGF188. Mech Dev 2002; 111(1-2):61-73.
19. Zelzer E, McLean W, Ng YS et al. Skeletal defects in VEGF(120/120) mice reveal multiple roles for VEGF in skeletogenesis. Development 2002; 129(8):1893-1904.

20. Mattot V, Moons L, Lupu F et al. Loss of the VEGF(164) and VEGF(188) isoforms impairs postnatal glomerular angiogenesis and renal arteriogenesis in mice. J Am Soc Nephrol 2002; 13(6):1548-1560.
21. Maes C, Stockmans I, Moermans K et al. Soluble VEGF isoforms are essential for establishing epiphyseal vascularization and regulating chondrocyte development and survival. J Clin Invest 2004; 113(2):188-199.
22. Gustafsson T, Ameln H, Fischer H et al. VEGF-A splice variants and related receptor expression in human skeletal muscle following submaximal exercise. J Appl Physiol 2005; 98(6):2137-2146.
23. Bates DO, Cui TG, Doughty JM et al. VEGF165b, an inhibitory splice variant of vascular endothelial growth factor, is down-regulated in renal cell carcinoma. Cancer Res 2002; 62(14):4123-4131.
24. Yang SP, Kwon BO, Gho YS et al. Specific interaction of VEGF165 with beta-amyloid, and its protective effect on beta-amyloid-induced neurotoxicity. J Neurochem 2005; 93(1):118-127.
25. Risau W. Embryonic angiogenesis factors. Pharmacol Ther 1991; 51(3):371-376.
26. Sato TN, Lougna S. Vasculogenesis and Angiogenesis. In: Tam JRaPPL, ed. Mouse Development: Patterning, Morphogenesis, and Organogenesis. Academic Press, 2002:211-228.
27. Speck N, Peeters M, Dzierzak E. Development of the vertebrate hematopoietic system. In: Tam JRaPPL, ed. Mouse Development: Patterning, Morphogenesis, and Organogenesis. Academic Press, 2002:191-206.
28. Harvey RP. Molecular determinants of cardiac development and congenital disease. In: Tam JRaPPL, ed. Mouse Development: Patterning, Morphogenesis, and Organogenesis. Academic Press, 2002:332-358.
29. Shalaby F, Rossant J, Yamaguchi TP et al. Failure of blood-island formation and vasculogenesis in Flk-1-deficient mice. Nature 1995; 376(6535):62-66.
30. Shalaby F, Ho J, Stanford WL et al. A requirement for Flk1 in primitive and definitive hematopoiesis and vasculogenesis. Cell 1997; 89(6):981-990.
31. Schuh AC, Faloon P, Hu QL et al. In vitro hematopoietic and endothelial potential of flk-1(-/-) embryonic stem cells and embryos. Proc Natl Acad Sci USA 1999; 96(5):2159-2164.
32. Fong GH, Rossant J, Gertsenstein M et al. Role of the Flt-1 receptor tyrosine kinase in regulating the assembly of vascular endothelium. Nature 1995; 376(6535):66-70.
33. Fong GH, Zhang L, Bryce DM et al. Increased hemangioblast commitment, not vascular disorganization, is the primary defect in flt-1 knock-out mice. Development 1999; 126(13):3015-3025.
34. Hiratsuka S, Minowa O, Kuno J et al. Flt-1 lacking the tyrosine kinase domain is sufficient for normal development and angiogenesis in mice. Proc Natl Acad Sci USA 1998; 95(16):9349-9354.
35. Hiratsuka S, Nakao K, Nakamura K et al. Membrane fixation of vascular endothelial growth factor receptor 1 ligand-binding domain is important for vasculogenesis and angiogenesis in mice. Mol Cell Biol 2005; 25(1):346-354.
36. Hiratsuka S, Kataoka Y, Nakao K et al. Vascular endothelial growth factor A (VEGF-A) is involved in guidance of VEGF receptor-positive cells to the anterior portion of early embryos. Mol Cell Biol 2005; 25(1):355-363.
37. Kearney JB, Kappas NC, Ellerstrom C et al. The VEGF receptor flt-1 (VEGFR-1) is a positive modulator of vascular sprout formation and branching morphogenesis. Blood 2004; 103(12):4527-4535.
38. Autiero M, Waltenberger J, Communi D et al. Role of PlGF in the intra- and intermolecular cross talk between the VEGF receptors Flt1 and Flk1. Nat Med 2003; 9(7):936-943.
39. Fujisawa H. From the discovery of neuropilin to the determination of its adhesion sites. Adv Exp Med Biol 2002; 515:1-12.
40. Soker S, Takashima S, Miao HQ et al. Neuropilin-1 is expressed by endothelial and tumor cells as an isoform-specific receptor for vascular endothelial growth factor. Cell 1998; 92(6):735-745.
41. Kawasaki T, Kitsukawa T, Bekku Y et al. A requirement for neuropilin-1 in embryonic vessel formation. Development 1999; 126(21):4895-4902.
42. Takashima S, Kitakaze M, Asakura M et al. Targeting of both mouse neuropilin-1 and neuropilin-2 genes severely impairs developmental yolk sac and embryonic angiogenesis. Proc Natl Acad Sci USA 2002; 99(6):3657-3662.
43. Minamisawa S, Gu Y, Ross Jr J et al. A post-transcriptional compensatory pathway in heterozygous ventricular myosin light chain 2-deficient mice results in lack of gene dosage effect during normal cardiac growth or hypertrophy. J Biol Chem 1999; 274(15):10066-10070.
44. Giordano FJ, Gerber HP, Williams SP et al. A cardiac myocyte vascular endothelial growth factor paracrine pathway is required to maintain cardiac function. Proc Natl Acad Sci USA 2001; 98(10):5780-5785.
45. Miquerol L, Langille BL, Nagy A. Embryonic development is disrupted by modest increases in vascular endothelial growth factor gene expression. Development 2000; 127(18):3941-3946.
46. Dor Y, Camenisch TD, Itin A et al. A novel role for VEGF in endocardial cushion formation and its potential contribution to congenital heart defects. Development 2001; 128(9):1531-1538.

47. Gitler AD, Lu MM, Epstein JA. PlexinD1 and semaphorin signaling are required in endothelial cells for cardiovascular development. Dev Cell 2004; 7(1):107-116.
48. Gu C, Rodriguez ER, Reimert DV et al. Neuropilin-1 conveys semaphorin and VEGF signaling during neural and cardiovascular development. Dev Cell 2003; 5(1):45-57.
49. Takahashi N, Seko Y, Noiri E et al. Vascular endothelial growth factor induces activation and subcellular translocation of focal adhesion kinase (p125FAK) in cultured rat cardiac myocytes. Circ Res 1999; 84(10):1194-1202.
50. Motoike T, Markham DW, Rossant J et al. Evidence for novel fate of Flk1+ progenitor: Contribution to muscle lineage. Genesis 2003; 35(3):153-159.
51. Choi K, Kennedy M, Kazarov A et al. A common precursor for hematopoietic and endothelial cells. Development 1998; 125(4):725-732.
52. Yamashita J, Itoh H, Hirashima M et al. Flk1-positive cells derived from embryonic stem cells serve as vascular progenitors. Nature 2000; 408(6808):92-96.
53. Ema M, Rossant J. Cell fate decisions in early blood vessel formation. Trends Cardiovasc Med 2003; 13(6):254-259.
54. Ishida A, Murray J, Saito Y et al. Expression of vascular endothelial growth factor receptors in smooth muscle cells. J Cell Physiol 2001; 188(3):359-368.
55. Gerber HP, Malik AK, Solar GP et al. VEGF regulates haematopoietic stem cell survival by an internal autocrine loop mechanism. Nature 2002; 417(6892):954-958.
56. Haruta H, Nagata Y, Todokoro K. Role of Flk-1 in mouse hematopoietic stem cells. FEBS Lett 2001; 507(1):45-48.
57. Hattori K, Dias S, Heissig B et al. Vascular endothelial growth factor and angiopoietin-1 stimulate postnatal hematopoiesis by recruitment of vasculogenic and hematopoietic stem cells. J Exp Med 2001; 193(9):1005-1014.
58. Ohm JE, Gabrilovich DI, Sempowski GD et al. VEGF inhibits T-cell development and may contribute to tumor-induced immune suppression. Blood 2003; 101(12):4878-4886.
59. Belteki G, Haigh J, Kabacs N et al. Conditional and inducible transgene expression in mice through the combinatorial use of Cremediated recombination and tetracycline induction. Nucleic Acids Res 2005; 33(5):e51.
60. Matsumoto K, Yoshitomi H, Rossant J et al. Liver organogenesis promoted by endothelial cells prior to vascular function. Science 2001; 294(5542):559-563.
61. Lammert E, Cleaver O, Melton D. Induction of pancreatic differentiation by signals from blood vessels. Science 2001; 294(5542):564-567.
62. Ankoma-Sey V, Matli M, Chang KB et al. Coordinated induction of VEGF receptors in mesenchymal cell types during rat hepatic wound healing. Oncogene 1998; 17(1):115-121.
63. Wong AK, Alfert M, Castrillon DH et al. Excessive tumor-elaborated VEGF and its neutralization define a lethal paraneoplastic syndrome. Proc Natl Acad Sci USA 2001; 98(13):7481-7486.
64. LeCouter J, Moritz DR, Li B et al. Angiogenesis-independent endothelial protection of liver: Role of VEGFR-1. Science 2003; 299(5608):890-893.
65. Inoue M, Hager JH, Ferrara N et al. VEGF-A has a critical, nonredundant role in angiogenic switching and pancreatic beta cell carcinogenesis. Cancer Cell 2002; 1(2):193-202.
66. Lammert E, Gu G, McLaughlin M et al. Role of VEGF-A in vascularization of pancreatic islets. Curr Biol 2003; 13(12):1070-1074.
67. Compernolle V, Brusselmans K, Acker T et al. Loss of HIF-2alpha and inhibition of VEGF impair fetal lung maturation, whereas treatment with VEGF prevents fatal respiratory distress in premature mice. Nat Med 2002; 8(7):702-710.
68. Oosthuyse B, Moons L, Storkebaum E et al. Deletion of the hypoxia-response element in the vascular endothelial growth factor promoter causes motor neuron degeneration. Nat Genet 2001; 28(2):131-138.
69. Zeng X, Wert SE, Federici R et al. VEGF enhances pulmonary vasculogenesis and disrupts lung morphogenesis in vivo. Dev Dyn 1998; 211(3):215-227.
70. Le Cras TD, Spitzmiller RE, Albertine KH et al. VEGF causes pulmonary hemorrhage, hemosiderosis, and air space enlargement in neonatal mice. Am J Physiol Lung Cell Mol Physiol 2004; 287(1):L134-142.
71. Lee CG, Link H, Baluk P et al. Vascular endothelial growth factor (VEGF) induces remodeling and enhances TH2-mediated sensitization and inflammation in the lung. Nat Med 2004; 10(10):1095-1103.
72. Robert B, Zhao X, Abrahamson DR. Coexpression of neuropilin-1, Flk1, and VEGF(164) in developing and mature mouse kidney glomeruli. Am J Physiol Renal Physiol 2000; 279(2):F275-282.
73. Eremina V, Sood M, Haigh J et al. Glomerular-specific alterations of VEGF-A expression lead to distinct congenital and acquired renal diseases. J Clin Invest 2003; 111(5):707-716.

74. Eremina V, Quaggin SE. The role of VEGF-A in glomerular development and function. Curr Opin Nephrol Hypertens 2004; 13(1):9-15.
75. Germani A, Di Carlo A, Mangoni A et al. Vascular endothelial growth factor modulates skeletal myoblast function. Am J Pathol 2003; 163(4):1417-1428.
76. van Weel V, Deckers MM, Grimbergen JM et al. Vascular endothelial growth factor overexpression in ischemic skeletal muscle enhances myoglobin expression in vivo. Circ Res 2004; 95(1):58-66.
77. Gerber HP, Vu TH, Ryan AM et al. VEGF couples hypertrophic cartilage remodeling, ossification and angiogenesis during endochondral bone formation. Nat Med 1999; 5(6):623-628.
78. Haigh JJ, Gerber HP, Ferrara N et al. Conditional inactivation of VEGF-A in areas of collagen2a1 expression results in embryonic lethality in the heterozygous state. Development 2000; 127(7):1445-1453.
79. Zelzer E, Mamluk R, Ferrara N et al. VEGFA is necessary for chondrocyte survival during bone development. Development 2004; 131(9):2161-2171.
80. Weninger W, Uthman A, Pammer J et al. Vascular endothelial growth factor production in normal epidermis and in benign and malignant epithelial skin tumors. Lab Invest 1996; 75(5):647-657.
81. Rossiter H, Barresi C, Pammer J et al. Loss of vascular endothelial growth factor a activity in murine epidermal keratinocytes delays wound healing and inhibits tumor formation. Cancer Res 2004; 64(10):3508-3516.
82. Larcher F, Murillas R, Bolontrade M et al. VEGF/VPF overexpression in skin of transgenic mice induces angiogenesis, vascular hyperpermeability and accelerated tumor development. Oncogene 1998; 17(3):303-311.
83. Xia YP, Li B, Hylton D et al. Transgenic delivery of VEGF to mouse skin leads to an inflammatory condition resembling human psoriasis. Blood 2003; 102(1):161-168.
84. Eichmann A, Makinen T, Alitalo K. Neural guidance molecules regulate vascular remodeling and vessel navigation. Genes Dev 2005; 19(9):1013-1021.
85. Carmeliet P. Blood vessels and nerves: Common signals, pathways and diseases. Nat Rev Genet 2003; 4(9):710-720.
86. Hobson MI, Green CJ, Terenghi G. VEGF enhances intraneural angiogenesis and improves nerve regeneration after axotomy. J Anat 2000; 197(Pt 4):591-605.
87. Jin K, Zhu Y, Sun Y et al. Vascular endothelial growth factor (VEGF) stimulates neurogenesis in vitro and in vivo. Proc Natl Acad Sci USA 2002; 99(18):11946-11950.
88. Sondell M, Lundborg G, Kanje M. Vascular endothelial growth factor has neurotrophic activity and stimulates axonal outgrowth, enhancing cell survival and Schwann cell proliferation in the peripheral nervous system. J Neurosci 1999; 19(14):5731-5740.
89. Matsuzaki H, Tamatani M, Yamaguchi A et al. Vascular endothelial growth factor rescues hippocampal neurons from glutamate-induced toxicity: Signal transduction cascades. FASEB J 2001; 15(7):1218-1220.
90. Damert A, Miquerol L, Gertsenstein M et al. Insufficient VEGFA activity in yolk sac endoderm compromises haematopoietic and endothelial differentiation. Development 2002; 129(8):1881-1892.
91. Haigh JJ, Morelli PI, Gerhardt H et al. Cortical and retinal defects caused by dosage-dependent reductions in VEGF-A paracrine signaling. Dev Biol 2003; 262(2):225-241.
92. Raab S, Beck H, Gaumann A et al. Impaired brain angiogenesis and neuronal apoptosis induced by conditional homozygous inactivation of vascular endothelial growth factor. Thromb Haemost 2004; 91(3):595-605.
93. Schwarz Q, Gu C, Fujisawa H et al. Vascular endothelial growth factor controls neuronal migration and cooperates with Sema3A to pattern distinct compartments of the facial nerve. Genes Dev 2004; 18(22):2822-2834.
94. Wada T, Haigh JJ, Ema M et al. Vascular endothelial growth factor directly inhibits primitive neural stem cell survival but promotes definitive neural stem cell survival. J Neurosci 2006; 26(25):6803-6812.
95. Cao L, Jiao X, Zuzga DS et al. VEGF links hippocampal activity with neurogenesis, learning and memory. Nat Genet 2004; 36(8):827-835.
96. Lambrechts D, Storkebaum E, Carmeliet P. VEGF: Necessary to prevent motoneuron degeneration, sufficient to treat ALS? Trends Mol Med 2004; 10(6):275-282.
97. Carmeliet P, Storkebaum E. Vascular and neuronal effects of VEGF in the nervous system: Implications for neurological disorders. Semin Cell Dev Biol 2002; 13(1):39-53.
98. Azzouz M, Ralph GS, Storkebaum E et al. VEGF delivery with retrogradely transported lentivector prolongs survival in a mouse ALS model. Nature 2004; 429(6990):413-417.
99. Storkebaum E, Lambrechts D, Dewerchin M et al. Treatment of motoneuron degeneration by intracerebroventricular delivery of VEGF in a rat model of ALS. Nat Neurosci 2005; 8(1):85-92.

CHAPTER 6

VEGF and Endothelial Guidance in Angiogenic Sprouting

Holger Gerhardt*

Abstract

The cellular actions of VEGF need to be coordinated to guide vascular patterning during sprouting angiogenesis. Individual endothelial tip cells lead and guide the blood vessel sprout, while neighbouring stalk cells proliferate and form the vascular lumen. Recent studies illustrate how endothelial DLL4/NOTCH signalling, stimulated by VEGF, regulates the sprouting response by limiting tip cell formation in the stalk. The spatial distribution of VEGF, in turn, regulates the shape of the ensuing sprout by directing tip cell migration and determining stalk cell proliferation.

Key Messages

- Angiogenesis is a guided process.
- Endothelial tip cells lead each vascular sprout.
- VEGF induces tip cell formation.
- VEGF gradients are formed by heparin-binding isoforms.
- VEGF gradients guide tip cell migration and gauge stalk cell proliferation.
- VEGF and NOTCH signalling cooperate to select and guide endothelial tip cells in retinal development.

Introduction

The term angiogenesis summarizes a set of morphogenic events that expand and fine-tune the initial, more primitive, embryonic vascular network into a hierarchical network of arterioles, venules and highly branched capillaries to provide efficient blood supply and organ specific vascular functions.[1] These "angiogenic" events include sprouting morphogenesis, intussuseptive growth, splitting, remodelling, stabilization and differentiation into arterioles, venules and capillaries. At the cellular level, angiogenesis involves at least two distinct cell types, endothelial cells and supporting mural cells (pericytes and vascular smooth muscle cells), and requires a number of different cellular functions, such as migration, proliferation, cell survival, differentiation and specialization. A plethora of factors are involved at different levels, either stimulating or inhibiting angiogenesis. However, vascular endothelial growth factor (VEGF or VEGFA) plays a key role in most, if not all morphogenic events during angiogenesis (reviewed in ref. 2).

More than a decade of research on VEGF and angiogenesis has provided evidence that VEGF has multiple roles in endothelial cells, controlling both physiological and pathological

*Holger Gerhardt: Vascular Biology Laboratory, Cancer Research UK— London Research Institute, Lincoln's Inn Fields Laboratories, 44 Lincoln's Inn Fields, London WC2A 3PX, UK. Email: holger.gerhardt@cancer.org.uk

VEGF in Development, edited by Christiana Ruhrberg. ©2008 Landes Bioscience and Springer Science+Business Media.

angiogenesis. A major challenge is therefore to clarify how the different cellular functions of VEGF are concerted into precise morphogenic events. For example, how are the proliferative and migratory responses of endothelial cells to VEGF integrated during sprouting angiogenesis to facilitate the protrusion of a new, diameter-controlled vascular tube? Why do endothelial cells in some instances proliferate, and in others migrate? What controls the direction of the migratory response to VEGF? Which cells are susceptible to regression in the absence of VEGF as a survival signal? How does VEGF stimulate arterial identity in some endothelial cells, while inducing fenestration in others? Clearly, with every new function of VEGF that is discovered, the number of new questions is steadily increasing. Possible scenarios that may explain the diversity of the cellular responses to VEGF include the presence of specific receptors or receptor/coreceptor pairs, a possible morphogen function with concentration-dependent effects, or the context of the endothelial cells in their microenvironment. In the present chapter, I will discuss recent experimental advances that explain how some of the cellular functions in response to VEGF are orchestrated to promote guided vascular sprouting.

Mechanics of Angiogenic Sprouting

In order to extend a new cellular tube from a preexisting (quiescent) vascular network, the endothelial cells must be coordinated in their response. Endothelial cells in a culture dish proliferate and migrate in response to VEGF stimulation. When wounded in a so-called "scratch assay", the monolayer will close again by both proliferation and migration. VEGF stimulation accelerates this process. A confluent monolayer of endothelial cells may in some aspects resemble a quiescent vessel wall. Unlike the closure of an endothelial (or epithelial) monolayer, the formation of a new sprout requires selection of a distinct site on the vessel where endothelial cells start to invade the surrounding tissue or matrix, whereas other cells along the vessel stay put (Fig. 1A). In theory, if all cells begin to migrate, the vessel should disintegrate. Conversely, if all cells were to proliferate, the vessel would likely only increase in diameter. Thus the first process in angiogenic sprouting must be the selection of a distinct site on the mother vessel where sprout formation is initiated. This selection process will have to be reiterated as the new sprout elongates, branches and connects with other sprouts to form an expanding network. Proliferation will need to occur to provide more cells for sustained sprouting. These basic principles may not only apply to angiogenic sprouting in vertebrates, but more generally to different types of tubular sprouting processes, whereever they occur throughout development in the animal kingdom. For example, tubular sprouting in the *Drosophila* trachea has been studied in great detail, and we can learn by comparison.

Cell specification is one of the fundamental principles during formation of the *Drosophila* tracheal system. Each sprout is headed by a specialized tip cell, the fate of which is controlled by a number of different signalling pathways containing DPP, NOTCH and FGFR (reviewed in ref. 3). In this system, the sprouting process is induced by FGF secreted from distinct cell clusters in the vicinity of the tracheal cells. The tip cells extend dynamic filopodia towards the FGF source and migrate in a directional fashion up the FGF gradients. In contrast, the following cells do not adopt the tip cell phenotype, but form the stalk during the sprouting process. Genetic studies and mosaic analysis have clarified that the tip cell fate is inhibited in stalk cells through signals that negatively regulate the FGFR signaling pathway (sprouty) as well as bi-directional signalling between tip and stalk through the DELTA/NOTCH pathway (reviewed in ref. 3). Overexpression of the FGFR in all tracheal cells results in ectopic filopodia extension from the stalk, indicating that FGFR levels and activity underly the distinction between tip and stalk cell behaviour and fate.[4,5]

In attempts to understand the process of angiogenic sprouting, early imaging of salamander tails in the 1930s provided the first notion of dynamic protrusive behaviour at the tip of each vascular sprout; subsequently, detailed images of vascular sprouts in the developing CNS identified elaborate filopodia extension from the sprout tip, leading the authors to speculate that specific tip cells, which extend these filopodia, function to read guidance cues in the tissue

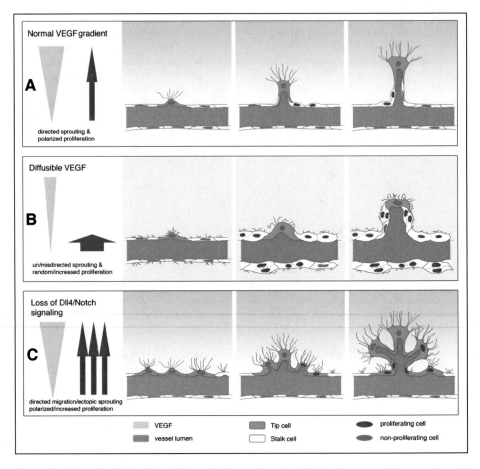

Figure 1. Schematic illustration of the cellular mechanisms that pattern vascular sprouting. A) Graded distribution of VEGF; sequential steps from left to right illustrate the induction of a tip cell (green) by VEGF (orange), polarization of a tip cell with rapid directed migration (blue arrow, left panel), and proliferation (pink nuclei) of the stalk cells (white). Polarization of the tip cell in a steep VEGF gradient leads to long, directed filopodia extension towards higher VEGF concentration. Polarized proliferation occurs with the division axis perpendicular to the long axis of the vessel. The pulling tip cell likely helps to polarize the stalk cell division. B) Diffuse distribution of VEGF, like in *Vegfa120/120* mice, leads to the undirected extension of short filopodia extension, although tip cell induction does occur. Tip cell migration is slow and many stalk cells proliferate due to widespread VEGF. Stalk cell proliferation is not polarized, causing vessel dilation/hypertrophy. C) Loss of DLL4/NOTCH signaling leads to excessive sprouting through increased tip cell numbers. NOTCH signaling normally inhibits the tip cell response in stalk cells. Tip cell numbers further increase through slightly elevated proliferation.

during the sprouting process (refs. 6-8, and references therein). Notably, these observations were made before VEGF was discovered,[9] and long before we learned how VEGF guides angiogenic sprouting by directing endothelial tip cell migration along extracellular gradients of heparin-binding isoforms.[10,11]

The ultrastructure of endothelial tip cells and their filopodia in contact with surrounding tissue was studied in the rat cerebral cortex in 1982.[7] The same team later used HRP injection and correlative microscopy to study tip cell filopodia and vascular lumen formation in regions where two sprouts connect, leading the authors to suggest that filopodia are important in anastomosis.[12] Tip cell filopodia where also found to lead sprouts and engage in sprout anastomosis in an angiogenesis model in vitro[13] and later also observed during sprouting angiogenesis in the chicken yolk sac[1] and CNS.[14,15] Risau and colleagues suggested through studies of the developing CNS that VEGF, expressed in the ventricular zone, may function to direct the vessel tips during the sprouting process.[15] Earlier, Ausprunk and Folkman had provided evidence that proliferation is largely confined to the stalk region during angiogenic sprouting.[16] Using irradiation to block proliferation, Folkman and colleagues also discovered that sprouting can in principle progress without cell division, indicating that the driving force for sprout elongation is likely a pulling force exerted by the tip cells, rather than a pushing force originating from dividing stalk cells. However, sustained sprouting beyond two days in the cornea pocket angiogenesis assay requires that cell numbers are replenished by division. Thus the mechanism of angiogenic sprouting involves (a) the local induction and selection of tip cells, (b) the directed migration of tip cells along a sufficiently adhesive substrate to provide the pulling force, and (c) the balanced proliferation of stalk cells during sprout elongation (Fig. 1A). Finally, the migration and explorative behaviour of the tip cells must cease upon anastomosis of two sprouts to establish a stable new connection between two branches of the growing vascular plexus.

Before I discuss how VEGF might affect each of these processes, I will briefly consider the composition and architecture of the tissue attracting the new vessel sprouts. Evidently, there will be tissue specific differences, and these are important for vascular patterning, as each organ brings its own requirement for efficient nutrient supply and oxygenation, but also creates physical constraints for blood vessels to restrain their growth and maintain adequate organ function. The postnatal retina of rodents provides an excellent model system to understand the interaction of tip and stalk cells with the surrounding tissue and to illustrate how VEGF expression, deposition and signalling controls guided vascular patterning within tissues.

The Mouse Retina Model

The retina comprises a sensory outpost of the brain, and as such may be regarded as a model for angiogenesis in the developing CNS in general. However, unlike most other parts of the CNS, the retina of rodents becomes vascularized only after birth (for recent comprehensive reviews of retinal vascular development and the retina as a model for angiogenesis research, see refs. 17,18). Vessels emerge from a capillary ring at the optic disk and sprout radially just underneath the inner limiting membrane, the vitreal surface of the retina. The sprouting vessels form an elaborate network that reaches the retinal periphery about one week after birth. The vessels are guided by a network of astrocytes, which forms only days before the vessels grow. Genetic manipulation of the astrocyte density results in a similar alteration of vascular density, illustrating the close relationship of astrocytes and vessels.[10,19] In fact, astrocytic network formation is a prerequisite for retinal angiogenesis at this stage. This is further supported by the fact that animals without retinal astrocytes also lack retinal vessels throughout evolution. The speed and accuracy of primary plexus formation along the astrocytic network is truly remarkable and requires tight regulation, as delayed or impaired retinal angiogenesis leads to an invasion of hyaloid vessels from the vitreous into the retina, and the ensuing abberant retinal vasculature may cause retinal scarring and detachment.

Detailed analysis of the physical association of the vascular sprouts with the underlying astrocytic network showed that the endothelial tip cells elongated their filopodia almost exclusively along the astrocyte surface.[10,20] Indeed, filopodia without astrocyte contact failed to stretch towards the retinal periphery in a directed fashion, suggesting that the astrocyte surface has adhesive properties. Dorrell and colleagues proposed that R-cadherin may mediate at least part of this adhesive function,[20] however, unequivocal evidence hereof is lacking so far. Rather, we have found that R-cadherin deficient mice show no defects in tip cell filopodia alignment or vascular development in the retina (H. Gerhardt, J. Hakansson and H. Semb, unpublished observations).

The astrocytes themselves appear to respond to endothelial tip cell contact by undergoing maturation and remodeling.[21] In the mature retinal network, the vascular surface is completely covered by astrocytic endfeet, and their inimate relationship makes an important contribution to the induction and maintenance of the blood retina barrier.[22,23] At the time when the initial vascular plexus has reached the periphery, a wave of secondary angiogenesis is initiated from veins and capillaries in their vicinity. The new sprouts elongate along the retinal radial glia (Mueller glia) to branch and anastomose into two consecutive plexuses in the inner and outer plexiforme layers.

VEGF Gradients Guide Tip Cell Migration

The extracellular distribution of growth factors is often tightly controlled to ensure spatial patterning of the appropriate tissue response. In angiogenesis, the extracellular distribution of VEGF is controlled at the level of transcription, isoform splicing, cell surface retention and likely uptake and degradation to result in extracellular gradient formation for proper vascular patterning.

Eli Keshet and colleagues studied the mouse retina and later the human retina in detail and discovered that astrocytes produce ample amounts of VEGF.[24,25] It was also in this tissue that Keshet first identified hypoxia-dependent regulation of VEGF expression. In situ hybridisation for VEGF mRNA reveals an intriguing picture, with VEGF expression confined largely to the avascular periphery throughout retinal development (see Chapter 3 by M. Fruttiger). The avascular periphery shrinks as the retinal plexus advances, and so does the zone of high VEGF expression.[21] Thus VEGF production is spatially graded and ideally suited to provide directionality to the migratory response of the endothelial tip cells. In addition, VEGF gradient formation is supported by the distinct retention properties of the different VEGF isoforms (for details on VEGF isoforms see Chapter 1 by Y.S. Ng).

The longer splice isoforms (VEGF164 and VEGF188 in the mouse) contain C-terminal basic amino acids sequences that interact with negatively charged heparan sulfate side chains on the cell surface or in the extracellular matrix. The shorter VEGF120 lacks this C-terminal retention motif, but still binds and activates the FLT1/VEGFR1 and KDR/VEGFR2 receptors. A detailed study of mice genetically engineered to produce only single isoforms, in lieu of several alternatively spliced isoforms, confirmed that the heparin-binding isoforms are secreted, but retained close to producing cells and thereby display a graded protein distribution around the site of production.[11] Accordingly, retinal astrocytes secrete and retain VEGF protein in their environment in a graded fashion. Intriguingly, endothelial tip cells express KDR mRNA in abundance and localize the receptor protein to both their cell body as well as their filopodia.[2,10] Thus, in analogy to axonal growth cones, which carry receptors during neuronal guidance, the endothelial tip cell filopodia with their receptors are ideally suited to perform the sensing function originally postulated by Marin-Padilla.[8] Indeed, we observed in a series of gain and loss of function studies that the extracellular VEGF distribution is essential for the length and orientation of tip cell filopodia in vivo (see below).

Endothelial tip cells are most prominently found in the periphery of the developing vascular plexus, where most of the new sprouts are forming. Tip cell formation and sprouting co-distribute with areas of highest VEGF concentration, suggesting that tip cell formation and filopodia protrusion may be induced by VEGF. Indeed, we observed induction of new tip cells and excessive filopodia formation on hyaloid vessels exposed to high VEGF levels in transgenic animals overexpressing individual VEGF isoforms from the lens crystallin promoter.[10] Conversely, sequestration of endogenous VEGF by intraocular injection of soluble FLT1 rapidly leads to loss of tip cell filopodia.[10] Also, neutralizing antibodies to KDR, but not FLT1 inhibit filopodia formation, indicating that VEGF mediates tip cell induction and filopodia extension via activation of KDR.[10] However, FLT1 may help to shape the extracellular VEGF gradient, as the soluble form of FLT1 (sFLT1) might act as a VEGF sink, keeping VEGF levels low close to the vessel stalk cells.[26]

Although all VEGF isoforms appear to induce filopodia formation, the effects on filopodia morphology and vascular patterning are strictly isoform specific. For example, Ruhrberg and colleagues found that mouse hindbrains expressing solely VEGF188 contained many tip cells that extended numerous long filopodia, whereas tip cells were sparser and tip cell filopodia shorter in the presence of VEGF120 only.[11] Correlating with the tip cell phenotype, VEGF120 induced the formation of a poorly branched network of vessels with enlarged diameter. Vice versa, mice producing only VEGF188 developed a highly branched network of thin vessels. Finally, mice producing only VEGF164, which is a heparin-binding isoform that is also diffusible, developed vessels that were indistinguishable from those in wild type littermates.

Close examination of the site of VEGF expression, and the localization of the VEGF protein, had demonstrated for the first time that VEGF120 diffused over considerable distance within tissues, whereas the heparin binding isoforms build a steep extracellular VEGF gradient. Ruhrberg and colleagues therefore concluded that diffusible VEGF120 reached the endothelium over large distances and stimulated continued proliferation of endothelial cells, which in turn increased vessel diameter, whereas VEGF188 gradients were so steep that vessels branched excessively.[2,11]

A close examination of vascular patterning in the retinas of *Vegfa120/120* mice illustrated that tip-cell migration and expansion of the retinal vascular plexus is dramatically reduced as a consequence of a shallow VEGF gradient.[10] Interestingly, the length of the tip cell filopodia closely correlated with the speed of retinal plexus migration and the assumed gradient in all experimental observations. Indeed, vessels in mice that carry only one *Vegfa120* allele instead of the correctly spliced wild type allele (*Vegfa+/120* mice), already show reduced tip cell migration, filopodia shortening and perturbed vessel branching.

Further mechanistic insight into the role of VEGF gradients in tip cell migration emerged through a series of experiments involving intraocular VEGF injection. Injection of the heparin-binding VEGF164 isoform rapidly reduced tip cell migration, shortened tip cell filopodia and enlarged stalk diameter, features that were highly reminiscent of the phenotype of mice expressing VEGF120 only. Taken together, these results suggested that the key parameter for tip cell migration, tip cell polarization and directional filopodia extension was the precisely controlled extracellular localization of VEGF.

RT-PCR revealed that the heparin-binding VEGF164 is the dominant VEGF isoform in the retina. Thus the normal endogenous distribution of VEGF is likely controlled by retention of this growth factor close to the site of production to build the extracellular VEGF gradient. This gradient is required for proper tip cell polarization, directed filopodia extension and migration, following the basic principles of chemotaxis (Figs. 1A,B).

Balancing Tip Cell Migration and Stalk Cell Migration through Extracellular VEGF Gradients

The proliferation of stalk cells is also controlled by VEGF distribution (Fig. 1). Here, the local availability of VEGF within the plexus appears to determine cell division. In a normal, unperturbed retina, only a limited zone of the peripheral vascular plexus behind the sprouting front shows significant endothelial cell proliferation, as judged by BrdU incorporation, Ki67 or phospho-histone 3 labelling. Injection of VEGF, however, leads to widespread proliferation throughout the vascular plexus. This clearly shows that most endothelial cells in the growing vascular plexus, irrespective of their differentiation status and smooth muscle coverage, can in principle respond to VEGF by proliferation. During normal development, the amount of VEGF produced correlates closely with local perfusion and oxygen supply, as *Vegfa* gene transcription is controlled by hypoxia. As mentioned above, the avascular periphery in the retina is hypoxic and produces very high levels of VEGF. Within the plexus, on the other hand, VEGF production is largely downregulated. However, local differences in perfusion and oxygenation result in a residual amount of VEGF production. Low oxygenation of venous blood correlates with higher VEGF production from astrocytes around the venous area, while astrocytes close to arteries produce little or no VEGF.[27] The distribution of endothelial proliferation follows this pattern exactly, with significant levels in veins and the surrounding plexus and essentially no proliferation in and around arteries.

Intriguingly, it is the relative difference in VEGF production around arteries and veins compared to the VEGF levels in front of the sprouting tip region that appears to regulate the patterning of the primary plexus ahead of the remodeling veins and arteries. Where the difference is greater, around the arterial regions, the implied long-range gradient of extracellular VEGF should be steep and result in the directed extension of long filopodia and rapid tip cell advancement over the retinal surface. Near veins, where this gradient is expected to be less steep, tip cell advancement and polarity should be comparably reduced. The opposite would be true for proliferation of the stalk cells, resulting low stalk cell proliferation in the former case, and higher stalk cell proliferation in the latter case. Taken together, these findings illustrate how the spatial VEGF distribution might balance and regulate the VEGF responses of tip and stalk cells and thus shapes the morphology of nascent vessels (Figs. 1A, B).

We were able to validate the concept that the spatial distibution of VEGF controls endothelial responses to VEGF further by hyperoxia and hypoxia treatment of neonatal mice. Exposure to lower than normal oxygen levels increased VEGF production and therefore raised the level of residual VEGF available to the vascular plexus, but left the avascular periphery unperturbed; in contrast, exposure to higher than normal oxygen levels had the opposite effect. Conceptually, high oxygen should thus steepen the gradient, whereas hypoxia should decrease it. Consistent with this idea, high oxygen levels enhancd tip cell migration and reduced stalk cell proliferation. Hypoxia in turn reduced tip cell migration and increased stalk cell proliferation, resulting in the formation of a vascular plexus that appeared very similar to the one observed in mice expressing VEGF120 only. Together, these data illustrate not only how important the extracellular distribution of VEGF is for vascular patterning, but also how the regulation of VEGF production and retention contributes to extracellular gradient formation to balance tip and stalk cell proliferation.

Recent data from Bautch and colleagues added another dimension to the control of stalk cell proliferation by VEGF distribution. Monitoring endothelial cell division in mouse embryonic stem cell-derived embryoid bodies, they found that stalk cells normally divided with their plane of cytokinesis perpendicular to the vessel long axis.[28] This polarized cell division promoted vessel elongation during sprouting. Interestingly, stalk cell polarization occured independently of flow, but was instead regulated by heparin-binding VEGF isoforms. Bautch and colleagues had earlier suggested that cell proliferation and VEGF gradient formation in the embryoid body system is regulated by FLT1, in particular the soluble form, which acts as an extracellular VEGF sink around established vessels.[26,29] They then showed that diffusible VEGF120 and loss of soluble FLT1 both led to randomization of the cell division axis, and thus resulted in abberant vessels with enlarged diameter.

Based on the results described above, it is tempting to speculate that the vascular abnormalities in mice expressing VEGF120 as the only VEGF isoform are caused by a combination of loss of tip cell polarization and increased stalk cell proliferation, as well as the loss of polarization of stalk cell cytokinesis (Fig. 1B).

Precisely heparin-binding isoforms polarize the division axis of stalk cells remains unclear. Many studies illustrate that endothelial cell shape, polarity and division axis can be regulated by flow through shear stress.[30] While this is highly unlikely to occur in the sprouting regions, where both flow and pressure are low, it is possible that the pulling force generated by the tip cells (see above) is transmitted to the stalk cells to polarize their division axis. Thus, the loss of tip cell polarization and directed migration in the absence of VEGF gradients could directly be responsible for a loss of stalk cell polarization. Further studies are required to investigate this link.

Tip Cell Formation following VEGF Stimulation Is Controlled by Dll4/NOTCH1 Signaling

The concept of endothelial tip and stalk cells has helped to direct current research towards identifying distinct cellular responses of different endothelial cell populations and their integration during angiogenic sprouting. Which signals select a tip cell from a given endothelial cell population, and what stops the stalk cells from responding in a similar fashion? Clearly both types of endothelial cells are stimulated by the same growth factor, VEGF, and both respond through KDR signaling, yet their behaviour is very different. Our initial characterization of the tip cell illustrated that tip and stalk cells carry a differential transcriptional signature. Today, we know only a handful of genes that are expressed preferentially in the tips, including *Kdr*, *Pdgfb* (encoding the patelet-derived growth factor B), *Apln* (encoding apelin), *Dll4* (encoding delta-like 4).

In vitro, VEGF induces a large number of genes in endothelial cells from various origins. One of the genes induced by VEGF is the NOTCH ligand DLL4. In situ hybridization suggested that *Dll4* expression is restricted to developing arteries and the tips of vascular sprouts.[31,32] A number of recent studies have now addressed the function of DLL4/NOTCH signaling in angiogenesis. Very similar to VEGF, DLL4 levels appear to require very tight regulation, as haploinsufficiency for *Dll4* causes embryonic lethality.[33-35] Interestingly, loss of DLL4/NOTCH signaling in angiogenesis assays in vitro[36] and in mouse and zebrafish in vivo leads to ectopic sprouting and increased tip cell numbers[37-41] (Fig. 1C). Monitoring NOTCH signaling activity through NOTCH target genes and a transgenic NOTCH reporter in the mouse retina, we observed strong NOTCH signaling in the sprouting zone. As observed for *Dll4* expression, NOTCH reporter activity displayed a "salt and pepper" distribution pattern among the endothelial cells in the sprouting zone. Intriguingly, *Dll4* mRNA was almost never observed to be strongly expressed in two neighboring cells at any given time, independent of their position at the tip or in the stalk of the sprouts. Loss of NOTCH activity recapitulates the phenotype caused by loss of DLL4, with strongly increased filopodia protrusions, increased tip cell numbers, excessive sprouting and fusion. These data suggested that DLL4/NOTCH signaling controls protrusive activity. The fact that VEGF induces *Dll4* expression,[39,42] and the observation of increased filopodia formation, sprouting and branching specifically in the region exposed to VEGF in mouse mutants together suggest that DLL4/NOTCH signalling functions to limit tip cell formation in the vascular zone exposed to VEGF.

DLL4, like other NOTCH ligands, activates the NOTCH receptor in a cell-cell contact dependent manner. A series of protease cleavage events of both the ligand and the receptor is required for signaling. Gamma secretase is the last cleavase during receptor activation, severing the intracellular NOTCH domain from the transmembrane region.[43,44] The NOTCH intracellular domain (NICD) is translocated to the nucleus, where it binds to RBPJ (previous names: RbpSuh and CBF1) to activate transcription of target genes, including members of the HES and HEY family of transcriptional repressors (reviewed in refs. 45,46). Short-term inhibition of gamma secretase in the retina rapidly increases filopodia formation, in particular from cells situated in the stalk, suggesting that NOTCH is required in stalk cells to suppress protrusive activity. Moreover, expression of the tip cell marker *Pdgfb* becomes widespread at the front, suggesting that the tip cell program is activated in stalk cells. Mosaic analysis in mouse retina and zebrafish showed that cells unable to receive NOTCH signaling are more likely to adopt a tip cell phenotype, whereas activation of NOTCH prohibits cells from becoming tip cells.[37,40] These data argue for a model in which individual cells that are stimulated by VEGF compete for the leadership position, i. e. the tip cell phenotype.

How the competitive advantage of one cell over the other is established remains unclear. Eichmann and coworkers suggested that KDR levels determine the tip cell response and are controlled by NOTCH signaling.[41] This hypothesis is indeed attractive, as a negative feedback loop in which VEGF induces DLL4, which in turn activates NOTCH in neighboring cell to suppress KDR levels and thus KDR activity, would suffice to pattern the endothelial population into tip and stalk cells. In the analogous case of the *Drosophila* trachea, FGFR levels (breathless) are indeed capable of selecting the tip cells in response to FGF (branchless) stimulation.[4,5] Alternative mechanisms are possible, in particular as coreceptors for VEGF may regulate the downstream activity and specificity of the KDR signaling pathway.

Importantly, excessive numbers of tip cells subsequent to DLL4 or NOTCH inhibition increase vessel sprouting and branching, and therefore also vascular density; however, the ensuing vascular network is poorly functional. Both in the retina, and in tumour models, the tissue becomes hypoxic and undersupplied, despite the increased vascular density. In fact, two papers from Regeneron and Genentech demonstrated that tumour growth is strongly reduced as a result of poor vascular function in *Dll4* heterozygous mutant mice.[47,48] Finally, Leslie and colleagues illustrated that DLL4/NOTCH signaling also functions to terminate tip cell protrusive acitivity once two tip cells fuse in the dorsal longitudinal anastomosing vessel of zebrafish.[38]

Conclusions

VEGF stimulates endothelial cells to sprout and proliferate to form new vessel structures. VEGF induces DLL4, which functions to pattern the endothelial population into tip and stalk cells. The tip cells then migrate along the VEGF gradients, whereas stalk cell proliferate in a polarized fashion to supply further endothelial cells. The balance between migration and polarized proliferation controls the length and diameter of the stalk (Figs. 1A-C). VEGF gradients, arising through regulated retention of VEGF on the cell surface and in the extracellular matrix, govern these polarization processes. In the future, it will be interesting to examine how matrix molecules participate in these mechanisms and how VEGF retention and DLL4/NOTCH signalling influence each other in the patterning process.

Acknowledgements

I wish to thank all colleagues who contributed to the recent advance in the understanding of guided vascular patterning in angiogenesis, including the authors of many papers not cited here due to space restriction. H.G. is supported by Cancer Research UK.

References

1. Risau W. Mechanisms of angiogenesis. Nature 1997; 386:671-674.
2. Ruhrberg C. Growing and shaping the vascular tree: Multiple roles for VEGF. Bioessays 2003; 25(11):1052-1060.
3. Metzger RJ, Krasnow MA. Genetic control of branching morphogenesis. Science 1999; 284(5420):1635-1639.
4. Ghabrial AS, Krasnow MA. Social interactions among epithelial cells during tracheal branching morphogenesis. Nature 2006; 441(7094):746-749.
5. Ribeiro C, Ebner A, Affolter M. In vivo imaging reveals different cellular functions for FGF and Dpp signaling in tracheal branching morphogenesis. Dev Cell 2002; 2(5):677-683.
6. Bär T, Wolff JR. The formation of capillary basement membranes during internal vascularization of the rat's cerebral cortex. Z Zellforsch 1972; 133:231-248.
7. Mato M, Ookawara S. Ultrastructural observation on the tips of growing vascular cords in the rat cerebral cortex. Experientia 1982; 38(4):499-501.
8. Marin-Padilla M. Early vascularization of the embryonic cerebral cortex: Golgi and electron microscopic studies. J Comp Neurol 1985; 241(2):237-249.
9. Leung DW, Cachianes G, Kuang WJ et al. Vascular endothelial growth factor is a secreted angiogenic mitogen. Science 1989; 246(4935):1306-1309.
10. Gerhardt H, Golding M, Fruttiger M et al. VEGF guides angiogenic sprouting utilizing endothelial tip cell filopodia. J Cell Biol 2003; 161(6):1163-1177.
11. Ruhrberg C, Gerhardt H, Golding M et al. Spatially restricted patterning cues provided by heparin-binding VEGF-A control blood vessel branching morphogenesis. Genes Dev 2002; 16(20):2684-2698.
12. Mato M, Ookawara S, Namiki T. Studies on the vasculogenesis in rat cerebral cortex. Anat Rec 1989; 224(3):355-364.

13. Flamme I, Baranowski A, Risau W. A new model of vasculogenesis and angiogenesis in vitro as compared with vascular growth in the avian area vasculosa. Anat Rec 1993; 237(1):49-57.
14. Kurz H, Gartner T, Eggli PS et al. First blood vessels in the avian neural tube are formed by a combination of dorsal angioblast immigration and ventral sprouting of endothelial cells. Dev Biol 1996; 173(1):133-147.
15. Breier G, Risau W. The role of vascular endothelial growth factor in blood vessel formation. Trends in Cell Biol 1996; 6:454-456.
16. Ausprunk DH, Folkman J. Migration and proliferation of endothelial cells in preformed and newly formed blood vessels during tumor angiogenesis. Microvasc Res 1977; 14(1):53-65.
17. Fruttiger M. Development of the retinal vasculature. Angiogenesis 2007; 10(2):77-88.
18. Uemura A, Kusuhara S, Katsuta H et al. Angiogenesis in the mouse retina: A model system for experimental manipulation. Exp Cell Res 2006; 312(5):676-683.
19. Fruttiger M, Calver AR, Kruger WH et al. PDGF mediates a neuron-astrocyte interaction in the developing retina. Neuron 1996; 17(6):1117-1131.
20. Dorrell MI, Aguilar E, Friedlander M. Retinal vascular development is mediated by endothelial filopodia, a preexisting astrocytic template and specific R-cadherin adhesion. Invest Ophthalmol Vis Sci 2002; 43(11):3500-3510.
21. West H, Richardson WD, Fruttiger M. Stabilization of the retinal vascular network by reciprocal feedback between blood vessels and astrocytes. Development 2005; 132(8):1855-1862.
22. Goldstein GW. Endothelial cell-astrocyte interactions: A cellular model of the blood-brain barrier. Ann NY Acad Sci 1988; 529:31-39.
23. Abbott NJ, Ronnback L, Hansson E. Astrocyte-endothelial interactions at the blood-brain barrier. Nat Rev Neurosci 2006; 7(1):41-53.
24. Provis JM, Leech J, Diaz CM et al. Development of the human retinal vasculature: Cellular relations and VEGF expression. Exp Eye Res 1997; 65:555-568.
25. Stone J, Itin A, Alon T et al. Development of retinal vasculature is mediated by hypoxia-induced vascular endothelial growtn factor (VEGF) expression by neuroglia. J Neurosci 1995; 15:4738-4747.
26. Kearney JB, Kappas NC, Ellerstrom C et al. The VEGF receptor flt-1 (VEGFR-1) is a positive modulator of vascular sprout formation and branching morphogenesis. Blood 2004; 103(12):4527-4535.
27. Claxton S, Fruttiger M. Oxygen modifies artery differentiation and network morphogenesis in the retinal vasculature. Dev Dyn 2005; 233(3):822-828.
28. Zeng G, Taylor SM, McColm JR et al. Orientation of endothelial cell division is regulated by VEGF signaling during blood vessel formation. Blood 2007; 109(4):1345-1352.
29. Kearney JB, Ambler CA, Monaco KA et al. Vascular endothelial growth factor receptor Flt-1 negatively regulates developmental blood vessel formation by modulating endothelial cell division. Blood 2002; 99(7):2397-2407.
30. McCue S, Dajnowiec D, Xu F et al. Shear stress regulates forward and reverse planar cell polarity of vascular endothelium in vivo and in vitro. Circ Res 2006; 98(7):939-946.
31. Shutter JR, Scully S, Fan W et al. Dll4, a novel Notch ligand expressed in arterial endothelium. Genes Dev 2000; 14(11):1313-1318.
32. Claxton S, Fruttiger M. Periodic Delta-like 4 expression in developing retinal arteries. Gene Expr Patterns 2004; 5(1):123-127.
33. Gale NW, Dominguez MG, Noguera I et al. Haploinsufficiency of delta-like 4 ligand results in embryonic lethality due to major defects in arterial and vascular development. Proc Natl Acad Sci USA 2004; 101(45):15949-15954.
34. Krebs LT, Shutter JR, Tanigaki K et al. Haploinsufficient lethality and formation of arteriovenous malformations in Notch pathway mutants. Genes Dev 2004; 18(20):2469-2473.
35. Duarte A, Hirashima M, Benedito R et al. Dosage-sensitive requirement for mouse Dll4 in artery development. Genes Dev 2004; 18(20):2474-2478.
36. Sainson RC, Aoto J, Nakatsu MN et al. Cell-autonomous notch signaling regulates endothelial cell branching and proliferation during vascular tubulogenesis. FASEB J 2005; 19(8):1027-1029.
37. Hellstrom M, Phng LK, Hofmann JJ et al. Dll4 signalling through Notch1 regulates formation of tip cells during angiogenesis. Nature 2007; 445(7129):776-780.
38. Leslie JD, Ariza-McNaughton L, Bermange AL et al. Endothelial signalling by the Notch ligand Delta-like 4 restricts angiogenesis. Development 2007; 134(5):839-844.
39. Lobov IB, Renard RA, Papadopoulos N et al. Delta-like ligand 4 (Dll4) is induced by VEGF as a negative regulator of angiogenic sprouting. Proc Natl Acad Sci USA 2007; 104(9):3219-3224.
40. Siekmann AF, Lawson ND. Notch signalling limits angiogenic cell behaviour in developing zebrafish arteries. Nature 2007; 445(7129):781-784.

41. Suchting S, Freitas C, le Noble F et al. The Notch ligand Delta-like 4 negatively regulates endothelial tip cell formation and vessel branching. Proc Natl Acad Sci USA 2007; 104(9):3225-3230.
42. Liu ZJ, Shirakawa T, Li Y et al. Regulation of Notch1 and Dll4 by vascular endothelial growth factor in arterial endothelial cells: Implications for modulating arteriogenesis and angiogenesis. Mol Cell Biol 2003; 23(1):14-25.
43. Berezovska O, Jack C, McLean P et al. Rapid Notch1 nuclear translocation after ligand binding depends on presenilin-associated gamma-secretase activity. Ann N Y Acad Sci 2000; 920:223-226.
44. De Strooper B, Annaert W, Cupers P et al. A presenilin-1-dependent gamma-secretase-like protease mediates release of Notch intracellular domain. Nature 1999; 398(6727):518-522.
45. Ehebauer M, Hayward P, Martinez-Arias A. Notch signaling pathway. Sci STKE 2006; 2006(364):cm7.
46. Iso T, Kedes L, Hamamori Y. HES and HERP families: Multiple effectors of the Notch signaling pathway. J Cell Physiol 2003; 194(3):237-255.
47. Noguera-Troise I, Daly C, Papadopoulos NJ et al. Blockade of Dll4 inhibits tumour growth by promoting non-productive angiogenesis. Nature 2006; 444(7122):1032-1037.
48. Ridgway J, Zhang G, Wu Y et al. Inhibition of Dll4 signalling inhibits tumour growth by deregulating angiogenesis. Nature 2006; 444(7122):1083-1087.

CHAPTER 7

Vascular and Nonvascular Roles of VEGF in Bone Development

Christa Maes* and Geert Carmeliet

Abstract

The majority of bones in the skeleton develop through the process of endochondral ossification. During this process, avascular cartilage becomes gradually replaced by highly vascularized bone tissue. VEGF is an essential mediator of all 3 key vascularization stages of endochondral bone development, and, in addition, exerts multiple nonvascular functions during each of these stages by acting directly upon the involved bone cells. In this chapter, we will discuss the various lines of evidence which demonstrate that the three major VEGF isoforms are essential to coordinate bone vascularization, cartilage morphogenesis and ossification during endochondral bone formation.

Key Messages
- VEGF is expressed by bone cells (osteoclasts, osteoblasts, and chondrocytes) involved in the process of endochondral ossification.
- VEGF receptors are expressed by endothelial cells and bone cells during endochondral ossification.
- VEGF controls the timely invasion of endothelial cells and osteoclasts/chondroclasts into developing long bones during primary ossification.
- VEGF regulates the proliferation, differentiation and/or survival of osteoclasts, osteoblasts and chondrocytes.
- The matrix-binding VEGF isoforms mediate metaphyseal angiogenesis and thereby regulate both trabecular bone formation and growth plate morphogenesis during endochondral bone formation.
- The soluble VEGF isoforms are required for epiphyseal vascularization and secondary ossification in growing long bones.

Introduction

The assembly of the skeleton during embryonic development relies on the formation of as many as 206 separate bones at sites distributed all over the body. Two distinct mechanisms are responsible for bone formation: intramembranous and endochondral ossification. During intramembranous ossification, bones develop directly from soft connective tissue. First, mesenchymal precursor cells aggregate at the site of the future bone formation, and they then differentiate into osteoblasts. The osteoblasts deposit bone matrix (osteoid) rich in type I

*Corresponding Author: Christa Maes—Laboratory for Experimental Medicine and Endocrinology, K. U. Leuven, Gasthuisberg, Herestraat 49, 3000 Leuven, Belgium. Email: christa.maes@med.kuleuven.be

VEGF in Development, edited by Christiana Ruhrberg. ©2008 Landes Bioscience and Springer Science+Business Media.

collagen, which later becomes mineralized. Terminally differentiated osteoblasts become entrapped in the bone as osteocytes. This type of bone deposition occurs in close spatial interaction with vascular tissue, but little is known about the role of VEGF in this process. Truly membranous bones are the flat bones of the skull (calvarial bones and mandibles) and parts of the clavicles. The long bones of the axial and appendicular skeleton also develop from mesenchymal condensations, but here these cells differentiate into chondrocytes to form a cartilaginous model of the future bone, the cartilage anlagen. This avascular cartilage subsequently becomes replaced by highly vascularized bone tissue through the process of endochondral ossification, which encompasses 3 key vascularization stages: (i) initial vascular invasion of the cartilage anlagen to establish the primary center of ossification (diaphysis) (Fig. 1A-C,G); (ii) capillary invasion at the growth plate (metaphysis) to mediate rapid bone lengthening (Fig. 1D-F,H); and (iii) vascularization of the cartilage ends (epiphysis) to initiate secondary ossification (Fig. 1D-F,I).

In the late 1990's several in vitro experiments established that VEGF and its receptors are expressed in specific bone cell types, and a strong regulation of VEGF expression by osteo-modulators was observed. These earliest findings suggested a possible role for VEGF in bone formation in vivo. However, a generalized mouse knock-out model of VEGF could not be employed to determine the role of VEGF in bone development, due to lethality of even heterozygous VEGF knock-out embryos at a stage preceding the onset of skeletal development. Therefore, alternative approaches had to be used to explore the physiological role of VEGF in bone development. The first evidence for an important role for VEGF in postnatal metaphyseal bone development was found in juvenile mice after administration of a soluble truncated chimeric VEGF receptor, which consists of the FLT1 extracellular domain fused to an IgG-Fc domain and sequesters VEGF protein with high affinity.[1] In this model, VEGF inactivation suppressed blood vessel invasion at the growth plate and concomitantly inhibited endochondral bone formation. Another strategy to block VEGF function whilst circumventing the early embryonic lethality of VEGF null mice entailed the *Cre/LoxP*-mediated conditional inactivation of the VEGF gene (*Vegfa*) in type II collagen expressing chondrocytes.[2,3] Finally, expression of only one of the major VEGF isoforms also rescued the embryonic lethality of VEGF null mice, and was therefore able to reveal the specific contributions of these isoforms to bone development.[4-6] Altogether, these models have exposed multiple essential roles of VEGF in the sequential stages of endochondral bone formation (Fig. 1), as will be discussed.

In mice, the *Vegfa* gene encodes 3 major alternatively spliced isoforms: VEGF120, VEGF164 and VEGF188 (see Chapter 1 by Y.S. Ng). VEGF120 has a low affinity for heparin and is considered to be a freely diffusible protein. In contrast, VEGF164 and even more so VEGF188 bind heparin with high affinity; this is thought to facilitate their binding to heparan sulfate-containing proteoglycans on the cell surface and in the extracellular matrix (ECM), from which they can be released by proteolytic enzymes such as matrix metalloproteinases (MMPs). All VEGF isoforms are capable of binding the VEGF receptor tyrosine kinases FLT1 and KDR. In contrast, VEGF164, but not VEGF120 has been shown to bind to NRP1 and NRP2.

In this chapter, we will describe how VEGF expression by several different bone cell types mediates a multitude of effects during endochondral ossification. In particular, we focus on the role of VEGF as an essential mediator of all 3 key vascularization stages during endochondral bone development, and describe how VEGF exerts multiple nonvascular functions during each of these stages by acting directly upon the involved bone cells. Moreover, we will discuss how the study of bone development in transgenic mice expressing solely VEGF120 (*Vegfa120/120*), VEGF164 (*Vegfa164/164*) or VEGF188 (*Vegfa188/188*) revealed differential requirements for the VEGF isoforms at different stages of bone formation.

Expression and Regulation of VEGF and VEGF Receptors in Bone Cell Types

VEGF and its receptors are expressed by several different bone cell types involved in endochondral ossification.

Chondrocytes

Chondrocytes in the cartilage template and later in the growth plate first proliferate, and then progressively differentiate into mature hypertrophic cells. Several autocrine and/or paracrine factors have been implicated in chondrocyte development, including parathyroid hormone related protein (PTHRP; now known as parathyroid hormone like peptide, PTHLH), indian hedgehog (IHH), bone morphogenetic proteins (BMPs) and fibroblast growth factors (FGFs). PTHRP and IHH form a negative feedback signaling pathway to control the pace of chondrocyte development in the growth plate. IHH also coordinates chondrocyte and osteoblast differentiation, together with the transcription factor RUNX2 (runt related transcription factor; also known as core binding factor 1, CBFA1).[7,8]

Hypertrophic chondrocytes, but not immature chondrocytes, consistently express high levels of VEGF in vivo. One factor that may control this VEGF expression is RUNX2.[9] By expressing VEGF and other angiogenic stimulators, hypertrophic cartilage becomes a target for capillary invasion and angiogenesis. In contrast, immature cartilage remains avascular due to the production of angiogenic inhibitors. As a result, the center of the developing epiphyseal growth plate becomes hypoxic,[10] but chondrocytes are well capable of surviving this challenge. For example, bovine articular chondrocytes are able to survive under oxygen tensions ranging from <0.1% to 20% for at least 7 days in vitro, with no evident differences in cell division or differentiation.[11] In response to this physiological hypoxia, immature chondrocytes in the center of the epiphysis upregulate VEGF. Accordingly, VEGF mRNA and protein levels are increased by hypoxia in cultured embryonic limbs and primary chondrocytes in vitro.[5,12] In general, hypoxia induces the expression of VEGF and other genes involved in angiogenesis and glucose metabolism via two transcriptional regulators, the hypoxia-inducible factors HIF1A and HIF2A (previously known as HIF1 alpha and HIF2 alpha; see Chapter 3 by M. Fruttiger). In agreement, inactivation of HIF1A in epiphyseal chondrocytes abolished the upregulation of VEGF in response to hypoxia in vitro.[12] Mice with inactivation of HIF1A in cartilage nevertheless showed increased VEGF expression in the epiphysis, suggesting that HIF2A and/or other factors may compensate for HIF1A loss or contribute to VEGF upregulation in vivo.[10] Studies on the VEGF receptor profile in chondrocytes showed that immature epiphyseal chondrocytes in vivo express the two VEGF isoform-specific receptors NRP1 and NRP2, but no detectable levels of FLT1 or KDR were found.[3,5] Expression of KDR has however been reported in some other cartilage types, such as the permanent thyroid cartilage of humans and cultured hypertrophic chondrocytes of chicken.[13,14]

Osteoblasts

Osteoblasts share a common mesenchymal precursor with chondrocytes, and specific regulatory factors direct the osteo-chondroprogenitors to either one of these lineages. As such, RUNX2 dominates the control of osteoblast differentiation. Mature osteoblasts produce bone matrix and abundantly express type I collagen. In a later differentiation stage, osteoblasts mineralize the osteoid and are typified by expression of osteocalcin.

Several groups reported that osteoblastic cells of mouse, rat or human origin express VEGF and its receptors, with highest expression levels being found at the late differentiation stages.[15,16] As observed in chondrocytes, VEGF production in osteoblasts is stimulated by hypoxia in a process that involves HIF1A and HIF2A.[17-19] Furthermore, VEGF expression in osteoblasts is also induced by several osteotropic factors, including BMPs, transforming growth factor beta (TGFB), prostaglandins, insulin-like growth factor 1 (IGF1), platelet-derived growth factor (PDGF), fibroblast growth factor 2 (FGF2) and 1alpha,25-dihydroxyvitamin D_3, but it is inhibited by bone catabolic factors such as glucocorticoids.

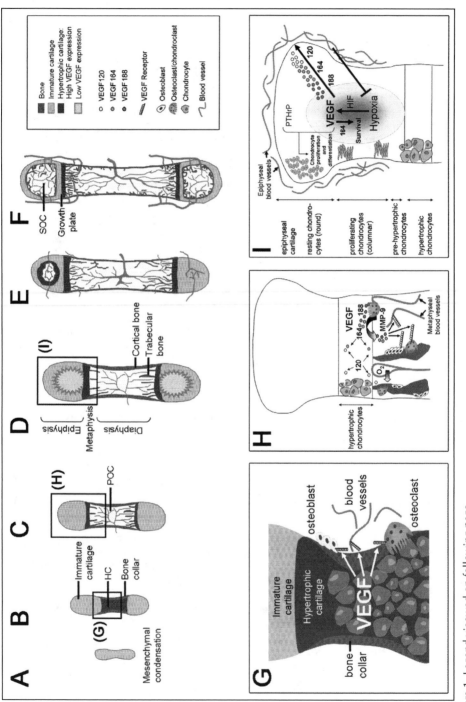

Figure 1. Legend viewed on following page.

Figure 1, viewed on previous page. Role of VEGF in endochondral bone development. A) Around E12 in mice, mesenchymal progenitor cells condense and differentiate into chondrocytes to form the cartilage anlagen that prefigure future bones. B) Around E14, hypertrophic chondrocytes (HC) form in the cartilage center (diaphysis), while cells in the connective tissue surrounding the cartilage (perichondrium) differentiate into osteoblasts. Osteoblasts deposit a mineralized bone matrix called bone collar around the cartilage. C) The primary ossification center (POC) forms when the diaphysis becomes vascularized and is invaded by osteoclasts, which resorb the hypertrophic cartilage, and by osteoblasts, which deposit bone matrix. The net result is replacement of the avascular cartilage anlage by a vascularized long bone. In the metaphysis, hypertrophic cartilage is continually replaced with trabecular bone, a process that relies on VEGF-mediated vascularization. D) Chondrocytes in the center of the avascular termini of the long bones (epiphyses) become hypoxic and express VEGF. E) Around postnatal day 5, epiphyseal vessels are attracted into the cartilage, likely in response to VEGF signals, and this initiates formation of the secondary ossification center (SOC). F) Discrete layers of residual chondrocytes form 'growth plates' between the epiphyseal and metaphyseal bone centers to support further postnatal longitudinal bone growth. G) Role of VEGF in orchestrating the development of the primary ossification center: VEGF is produced at high levels by hypertrophic chondrocytes, likely under the control of RUNX2. Endothelial cells, osteoblasts and osteoclasts express VEGF receptors and accumulate in the perichondrium. VEGF induces vascular invasion of the cartilage and may affect the differentiation and function of osteoblasts and osteoclasts. H) Metaphyseal vascularization, cartilage resorption and bone formation are coordinated by VEGF and MMP activity: Hypertrophic chondrocytes express VEGF120, VEGF164 and VEGF188; VEGF164 and VEGF188 are sequestered in the cartilage matrix, but can be released by proteases such as MMP9 to recruit endothelial cells and act upon osteoclasts and osteoblasts; MMP9 also supports cartilage resorption. I) Role of VEGF isoforms in epiphyseal bone development: When the avascular epiphyseal cartilage exceeds a critical size, immature chondrocytes in the center become hypoxic and express VEGF. The soluble VEGF isoforms VEGF120 and VEGF164 diffuse to the periphery to stimulate expansion of the epiphyseal vascular network and its subsequent invasion into the epiphysis at the start of secondary ossification. VEGF164 also promotes survival of hypoxic chondrocytes, and, in conjunction with other factors such as PTHRP, regulates chondrocyte development.

Osteoclasts

Osteoclasts are large, multinucleated cells that are endowed with the unique capacity to degrade mineralized tissues, a process in which secreted MMPs play an important role.[20] The development of osteoclasts is a complex multi-step process that involves at least two crucial signaling molecules expressed by osteoblasts and osteoblast progenitors: macrophage-colony stimulating factor (CSF1; previously known as M-CSF) and receptor activator of nuclear factor kappa B ligand (RANKL; now also known as tumor necrosis factor superfamily member 11, TNFSF11).[21]

Osteoclasts share a common hematopoietic precursor with monocytes and macrophages, and like them express FLT1 as their main VEGF receptor.[22,23] However, expression of KDR by cultured osteoclasts has also been reported.[24,25] Primary cultures of osteoclasts prepared from murine bone marrow were found to express VEGF by RT-PCR,[25] but these results have to be confirmed by additional experimental approaches, as these cultures also contained other bone marrow derived cells.

VEGF Is Required for the Formation of the Primary Ossification Center

Embryonic long bones first develop as avascular cartilage anlagen, but following the formation of a bone collar around the cartilage, vascular invasion takes place. Concomitant with vascular invasion, the hypertrophic cartilage matrix is degraded by invading osteoclasts and/or chondroclasts, and osteoblasts and marrow cells start to populate the primary ossification center. Blocking physically the vascular invasion of hypertrophic cartilage in embryonic day (E)14 skeletal explants halts bone development, indicating that the development of cartilage anlagen into proper long bones depends on the invasion of endothelial cells.[26] The vasculature is not only critical to supply oxygen, nutrients and growth enhancing molecules, but is also considered to be a major source of progenitors for the specific cell types that form bone and marrow. Several lines of evidence indicate that the timely invasion of endothelial cells and osteoclasts/chondroclasts during early bone development is dependent on VEGF, particularly VEGF164, through its direct actions on both endothelial cells and bone cells (Fig. 1G).

VEGF Controls the Initial Vascular Invasion during Formation of the Primary Ossification Center

At the time when the primary ossification center develops, VEGF is produced by perichondrial cells, possibly osteoblasts, and by diaphyseal hypertrophic chondrocytes (see below).[4,27] *Vegfa* transcription is thought to be induced by RUNX2, given that expression of VEGF and its receptors is impaired in the bones of RUNX2-deficient mice. Moreover, these mice show no vascular invasion into any skeletal element, consistent with the idea that VEGF is a critical vascular growth factor during bone formation.[9] This idea is particularly supported by the observation that the formation of the primary ossification center is delayed in cartilage explants cultured with a VEGF-inhibiting soluble chimeric FLT1 protein.[26] The initial vascular invasion and the formation of the primary ossification center are also delayed in *Vegfa120/120* and *Vegfa188/188* mice, but not in *Vegfa164/164* mice, suggesting that specifically the VEGF164 isoform is needed for this process.[4,5] This could be due to its interaction with NRP1. Alternatively, not the VEGF164 isoform in particular, but rather a combination of soluble and bound VEGF molecules may be needed to coordinate the initial capillary invasion into the cartilage anlagen, perhaps to form a chemoattractive gradient for ingrowing vessels. A similar mechanism has been proposed for angiogenesis in other organs (see Chapter 6 by H. Gerhardt). VEGF is also a chemoattractant for osteoclasts invading into developing bones. This process involves MMPs, raising the possibility that release of matrix-bound VEGF from the hypertrophic cartilage matrix by the action of MMPs may account for the close association of vascular and osteoclastic invasion.[28-30]

Nonvascular Roles of VEGF during the Formation of the Primary Ossification Center

VEGF appears to have also nonvascular roles in the initiation of bone development (Table 1), as supported by several observations: Firstly, ossification is reduced in embryonic metatarsals cultured with a soluble FLT1 chimera that inhibits VEGF.[4] Secondly, bone collar formation and cartilage calcification are decreased in embryonic *Vegfa120/120* and *Vegfa188/188* bones at a stage preceding vascularization.[4,5] Moreover, the analysis of *Vegfa*[120/20] bones revealed retarded terminal differentiation of hypertrophic chondrocytes and reduced expression of several markers for osteoblast and chondrocyte differentiation. Thirdly, the various bone cell types involved all express VEGF receptors: Hypertrophic chondrocytes express NRP1, perichondrial cells in vivo as well as osteoblastic cells in vitro express NRP1, NRP2, FLT1 and KDR, and osteoclasts express FLT1.[23,27,31] Although the precise effect of VEGF in vivo on the cell types involved in the initial stages of bone development is not yet fully understood, these findings suggest that VEGF isoforms take part in the timely differentiation of osteoblasts and chondrocytes. Interestingly, the expression of VEGF and NRP1 is already detected in the limb bud mesenchyme at E10.5 and at the periphery of the (pre)cartilage anlagen at E12.5.[2,32] Although unresolved at present, these data do raise the possibility that VEGF may function at even earlier stages in bone development, at the time when the cartilage condensations form.

VEGF Is Required for Metaphyseal Bone Development and Longitudinal Bone Growth

Longitudinal bone growth is mediated largely by the events occurring at the metaphyseal growth plate. The tight coupling between metaphyseal vascularization and endochondral bone development may be explained by the ability of blood vessels to function as a conduit which (i) allows cell types essential for bone morphogenesis, i.e., osteoclasts and osteoblasts, to migrate to the growth plate; (ii) removes end products of the resorption process; and (iii) supplies cells in the developing bone with oxygen, nutrients and growth factors/hormones required for their activity. Metaphyseal angiogenesis is induced by the matrix-binding VEGF isoforms and is an essential prerequisite for trabecular bone formation and growth plate morphogenesis. In addition, VEGF has been shown to directly affect osteoblasts and osteoclasts (Fig. 1H; Table 1).

The Role of VEGF in Metaphyseal Vascularization

During longitudinal bone growth, it is of utmost importance that the key mechanisms of endochondral ossification are rigorously coordinated, and the analysis of several different mouse models has demonstrated that VEGF-mediated metaphyseal angiogenesis plays a critical role in this process. Disruption of VEGF function in developing bones has been achieved by injection

---

Table 1. VEGF effects during the three key stages of endochondral bone development

VEGF Effects	POC Development	Longitudinal Bone Growth	SOC Development
Vascular	Vascular invasion	Metaphyseal vascularization	Epiphyseal vascularization
Nonvascular	HC differentiation (*)	HC apoptosis and resorption (*)	Chondrocyte proliferation
	OB development/ activity	OB and OC activity	and differentiation
	OC recruitment/activity		Chondrocyte survival

(*) The direct effect of VEGF in these processes has not unambiguously been shown in vitro.
Abbreviations: POC: primary ossification center; SOC: secondary ossification center; HC: hypertrophic chondrocyte; OB: osteoblast; OC: osteoclast.

of a soluble truncated chimeric VEGF receptor,[1] by targeted inactivation of the VEGF-isoforms VEGF164 and VEGF188 leaving only expression of the soluble isoform VEGF120,[4,6] and by conditional deletion of a single *Vegfa* allele in cells expressing type II collagen.[2] In all three instances, (partial) loss of VEGF function impaired metaphyseal bone vascularization. Specifically, vascularization was decreased and disorganized near the growth plate and, concomitantly, trabecular bone formation and bone growth were impaired. Typically, the hypertrophic chondrocyte zone of the growth plate was enlarged, due to reduced resorption and/or apoptosis. Thus, in metaphyseal bone development, VEGF functions to attract vessels to the growth plate, which is accompanied by hypertrophic chondrocyte apoptosis, cartilage resorption by osteoclasts/chondroclasts, and trabecular bone formation by osteoblasts.

Remarkably, a similar bone phenotype characterized by an enlarged hypertrophic chondrocyte zone was observed in mice deficient in MMP9 and/or MMP13.[33-35] This led to the hypothesis that MMP9 is produced by osteoclasts/chondroclasts to release ECM-bound VEGF from the chondrocyte matrix and thereby attract blood vessels (and more resorptive cells) to the growth plate (Fig. 1H).[8] In support of this model, *Vegfa164/164* and *Vegfa188/188* mice have no enlarged hypertrophic zone, nor do they display any other metaphyseal defect.[5] Thus, expression of either of the matrix-binding isoforms, VEGF164 or VEGF188, is necessary and sufficient to provide the signals required for normal metaphyseal vessel invasion and endochondral ossification (Fig. 1H). This observation suggests that the controlled VEGF release from the cartilage matrix favors organized directional angiogenesis, most likely by creating a VEGF gradient (see Chapter 6 by H. Gerhardt).[36] Alternatively, or additionally, VEGF signaling through NRP1 may be required as the impaired metaphyseal development is seen exclusively in *Vegfa120/120* mice and VEGF120 does not bind this receptor. Mice deficient in NRP1 die before the onset of bone development due to cardiovascular defects, but the analysis of conditional knockout mice with NRP1 inactivation exclusively in cartilage or bone may reveal a role for VEGF isoform signaling through NRP1 in bone cells.

Nonvascular Roles of VEGF during Longitudinal Bone Growth

Recent studies have suggested that VEGF may influence bone formation by directly affecting osteoblasts, as VEGF stimulates osteoblast differentiation and migration in vitro.[15,31,37] Moreover, adenovirus-mediated VEGF gene transfer induces bone formation by increasing osteoblast number and osteoid forming activity in vivo.[38] VEGF signaling has also been implicated in osteoclastogenesis and subsequent cartilage/bone resorption: Firstly, VEGF directly enhances the resorption activity and survival of mature osteoclasts in vitro.[24] Secondly, RANKL induces osteoclast differentiation from spleen- or bone marrow-derived precursors in culture when provided in combination with either VEGF or CSF1. Thirdly, VEGF, like CSF1, rescues osteoclast recruitment, survival and activity in osteopetrotic *op/op* mice, which carry an inactivating point mutation in the *Csf1* gene that results in low numbers of macrophages and a complete lack of mature osteoclasts.[23] Like monocytes and macrophages, osteoclasts predominantly express FLT1 rather than KDR (see above).[22,23] Moreover, FLT1 ligands are chemoattractive for both monocytes and osteoclasts.[29,39]

VEGF Affects Epiphyseal Cartilage Development and Formation of the Secondary Ossification Center

Because developing epiphyseal cartilage is avascular, its oxygenation is critically dependent on the vascular network overlying the cartilaginous surface, which is derived mainly from the epiphyseal arteries. These peripheral vessels later invade the cartilage to initiate the development of the secondary ossification center (Fig. 1D-F).[40] Recent studies have implicated the soluble VEGF isoforms in epiphyseal vascularization and secondary ossification.[5] In addition, VEGF was shown to act as a survival factor for chondrocytes in the hypoxic epiphysis (Fig. 1I; Table 1).[3,5]

The Role of VEGF in Epiphyseal Vascularization and the Initiation of Secondary Ossification

Vegfa188/188 mice form an abnormal capillary network overlying the epiphyses, a defect that is associated with increased hypoxia and massive apoptotic cell death in the interior of the cartilage. Chondrocytes located in the adjacent peripheral areas display an imbalance in their proliferation/differentiation rate. Impaired epiphyseal vascularization and chondrocyte development are most likely the cause of the strongly reduced long bone growth in *Vegfa188/ 188* mice, which display a dwarfed phenotype. Thus, the soluble VEGF isoforms are essential for epiphyseal vascularization, epiphyseal cartilage development and formation of the secondary ossification center.[5] Based on these findings, we suggest a model in which the progressive growth of the avascular epiphyseal cartilage results in a state of increased hypoxia that upregulates VEGF expression (Fig. 1I); soluble VEGF isoforms then diffuse from the hypoxic center towards the periphery to induce epiphyseal vessel outgrowth and thus reduce hypoxic stress. Subsequently, VEGF induces invasion of vessels into the cartilage to initiate secondary ossification. The presence of only VEGF188 is insufficient to stimulate epiphyseal vascularization, probably because this isoform binds tightly to matrix components and cellular surfaces, thereby failing to diffuse towards the periphery. Alternatively, the phenotype could be due to reduced VEGF signaling through NRP1, as it is not presently known if VEGF188 binds NRP1. However, this latter hypothesis seems less likely, since mice expressing only VEGF120 show normal epiphyseal vascularization, even though VEGF120 does not bind NRP1.[5] Epiphyseal vascular invasion and the subsequent development of the secondary ossification center are also impaired in mice lacking MT1-MMP,[41,42] suggesting that vascular invasion depends on both the degradation of the matrix by MT1-MMP and the attraction of blood vessels by VEGF. Interestingly, MT1-MMP upregulates VEGF expression in human breast carcinoma MCF7 cells,[43] but whether it also influences VEGF expression in cartilage is currently unknown.

Nonvascular Roles of VEGF in Epiphyseal Chondrocyte Development and Survival

In addition to its effects on epiphyseal vascularization, VEGF also directly affects chondrocyte development and survival in hypoxic cartilage. Firstly, VEGF is likely to act together with other factors, such as the PTHRP pathway (see above), to regulate the balance of chondrocyte proliferation and differentiation in the epiphysis (Fig. 1I).[5] This activity may be due to VEGF isoform signaling through NRP1, as this receptor is expressed on epiphyseal chondrocytes.[3,5] Secondly, both *Vegfa188/188* mice and mice with a complete inactivation of VEGF specifically in type II collagen-expressing cells show aberrant chondrocyte death, a phenotype similar to that seen in mice lacking HIF1A in cartilage.[3,5,10] In vitro cultures of *Vegfa 188/188* embryonic limbs revealed that expression of VEGF188 is not sufficient to protect chondrocytes against hypoxia-induced apoptosis, but supplementing recombinant VEGF164 rescued this defect. Thus, VEGF164 acts as a survival factor for hypoxic chondrocytes, possibly downstream of HIF1A.[3,5,10] The role of HIF1A in cartilage survival may therefore be at least in part due to its ability to upregulate VEGF, combined with the induction of anaerobic glycolytic metabolism.[44]

Conclusions and Future Perspectives

In this chapter, we have described the multiple essential roles that VEGF fulfills to support skeletal development. However, many mechanistic aspects of VEGF function in bone development remain to be elucidated, and the potential contribution of reduced VEGF signaling in bones to human disease has not yet been examined.

Novel Mouse Models to Understand VEGF Signaling in Bone Development

Whilst it is now evident that VEGF is essential to drive vascularization during endochondral bone development, the in vivo studies performed have also underscored our limited knowledge of the bone's vascular system itself and of its role in regulating the behavior of bone cells. For example, we don't know much about the types of blood vessels involved (capillaries, venous sinusoids, arteries), nor about the presence and role of pericytes or other peri-vascular cells. It is also still largely unclear how epiphyseal vascularization and the formation of vascular canals relate to secondary ossification. Moreover, we need to understand better the resorption processes that accompany vascular invasion. For instance, the contribution of specific proteases that release VEGF from the matrix at particular stages of endochondral bone development could be further addressed by the analysis of mice with specific mutations in MMP genes. The role of the VEGF-responsive cell types involved, such as endothelial cells and (other) resorbing cells, could be addressed by specifically targeting VEGF receptors.

Importantly, it has become clear that VEGF also exerts direct effects on several key bone cell types during endochondral bone development. These direct effects are still incompletely understood, for two reasons: Firstly, in vivo models have often been difficult to analyze, as the alteration of the VEGF expression levels or the VEGF isoform balance almost inevitably causes angiogenic defects, which in turn affect bone development. Secondly, in vitro models are limited in their capacity to accurately reproduce the complex differentiation processes that occur in vivo. However, the combination of both approaches, together with the use of transgenic mice carrying cell type-specific, temporally restricted or even cellular differentiation stage-specific knockout alleles for VEGF and its receptors, including the isoform-specific VEGF receptors, will provide this critical information. Aspects to be addressed include the precise effects, mechanisms of action and regulation of VEGF in osteoblasts, osteoclasts, and chondrocytes. Particularly regarding the regulation of VEGF in cartilage, much remains to be learned about the role of the hypoxia regulatory pathway. Inactivation of von Hippel Lindau (VHL), a mediator of HIF degradation, in murine cartilage was recently shown to alter chondrocyte proliferation, further underscoring that components of the hypoxia regulatory pathway play important physiological roles in cartilage, either directly and/or by affecting VEGF levels.[45] Furthermore, it will be interesting to see whether this pathway also plays a role in other bone cells. Finally, it will be exciting to explore if VEGF contributes to the very early stages of skeletal development, prior to vascular invasion and ossification, and whether VEGF regulates the development of joints.

From Mice to Men: A Role for VEGF Misexpression in Growth Disorders?

Animal studies have shown that the precise level of VEGF is critically important for embryonic development. Interestingly, the phenotypes of *Vegfa188/188* and *Vegfa120/120* mice suggest that altering the relative levels of the VEGF isoforms—without affecting the total level of VEGF—impairs developmental processes such as vascular network formation and skeletal development and growth. In particular, normal levels of the VEGF164 isoform appear to be critical for normal bone development in mice. However, it has not yet been examined if subtle variations in VEGF or VEGF isoform expression levels affect the development of the human skeleton. For example, it is conceivable that allelic variations in the human *VEGFA* gene promoter or abnormal *VEGFA* mRNA splicing affect either VEGF expression or the production of the VEGF165 isoform (the human ortholog of murine VEGF164). Hypothetically, such changes might cause or increase the risk of growth defects, or add to the severity of skeletal disorders caused by mutation in other genes (e.g., FGF receptor 3 or PTH/PTHRP receptor). It will be particularly important to examine if *VEGFA* gene polymorphisms are linked to the pathogenesis of human dwarfing syndromes or other skeletal pathologies, because low VEGF levels are already known to predispose humans and mice to motor neuron degeneration (see Chapter 8 by J. Krum, J. Rosenstein and C. Ruhrberg), and loss of VEGF164 expression acts as a modifier of DiGeorge syndrome, a disease with vascular and craniofacial abnormalities.[46,47] Studying the role of VEGF in bone development may also provide the basis for new therapies aimed at treating debilitating and/or unmanageable bone diseases, such as osteoporosis, bone metastases, and nonhealing fractures.

Acknowledgements

Christa Maes is a postdoctoral fellow of the Fund for Scientific Research Flanders (FWO).

References

1. Gerber HP, Vu TH, Ryan AM et al. VEGF couples hypertrophic cartilage remodeling, ossification and angiogenesis during endochondral bone formation. Nat Med 1999; 5:623-8.
2. Haigh JJ, Gerber HP, Ferrara N et al. Conditional inactivation of VEGF-A in areas of collagen2a1 expression results in embryonic lethality in the heterozygous state. Development 2000; 127:1445-1453.
3. Zelzer E, Mamluk R, Ferrara N et al. VEGFA is necessary for chondrocyte survival during bone development. Development 2004; 131:2161-2171.
4. Maes C, Carmeliet P, Moermans K et al. Impaired angiogenesis and endochondral bone formation in mice lacking the vascular endothelial growth factor isoforms VEGF164 and VEGF188. Mech Dev 2002; 111:61-73.
5. Maes C, Stockmans I, Moermans K et al. Soluble VEGF isoforms are essential for establishing epiphyseal vascularization and regulating chondrocyte development and survival. J Clin Invest 2004; 113:188-199.
6. Zelzer E, McLean W, Ng YS et al. Skeletal defects in VEGF(120/120) mice reveal multiple roles for VEGF in skeletogenesis. Development 2002; 129:1893-1904.
7. Kronenberg HM. Developmental regulation of the growth plate. Nature 2003; 423:332-336.
8. Karsenty G, Wagner EF. Reaching a genetic and molecular understanding of skeletal development. Dev Cell 2002; 2:389-406.
9. Zelzer E, Glotzer DJ, Hartmann C et al. Tissue specific regulation of VEGF expression during bone development requires Cbfa1/Runx2. Mech Dev 2001; 106:97-106.
10. Schipani E, Ryan HE, Didrickson S et al. Hypoxia in cartilage: HIF-1alpha is essential for chondrocyte growth arrest and survival. Genes Dev 2001; 15:2865-2876.
11. Grimshaw MJ, Mason RM. Bovine articular chondrocyte function in vitro depends upon oxygen tension. Osteoarthritis Cartilage 2000; 8:386-392.
12. Cramer T, Schipani E, Johnson RS et al. Expression of VEGF isoforms by epiphyseal chondrocytes during low-oxygen tension is HIF-1 alpha dependent. Osteoarthritis Cartilage 2004; 12:433-439.
13. Pufe T, Mentlein R, Tsokos M et al. VEGF expression in adult permanent thyroid cartilage: Implications for lack of cartilage ossification. Bone 2004; 35:543-552.
14. Carlevaro MF, Cermelli S, Cancedda R et al. Vascular endothelial growth factor (VEGF) in cartilage neovascularization and chondrocyte differentiation: Auto-paracrine role during endochondral bone formation. J Cell Sci 2000; 113:59-69.
15. Deckers MML, Karperien M, van der Bent C et al. Expression of vascular endothelial growth factors and their receptors during osteoblast differentiation. Endocrinology 2000; 141:1667-1674.
16. Harper J, Gerstenfeld LC, Klagsbrun M. Neuropilin-1 expression in osteogenic cells: Down-regulation during differentiation of osteoblasts into osteocytes. J Cell Biochem 2001; 81:82-92.
17. Akeno N, Czyzyk-Krzeska MF, Gross TS et al. Hypoxia induces vascular endothelial growth factor gene transcription in human osteoblast-like cells through the hypoxia inducible factor-2alpha. Endocrinology 2001; 142:959-962.
18. Kim HH, Lee SE, Chung WJ et al. Stabilization of hypoxia-inducible factor-1alpha is involved in the hypoxic stimuli-induced expression of vascular endothelial growth factor in osteoblastic cells. Cytokine 2002; 17:14-27.
19. Steinbrech DS, Mehrara BJ, Saadeh PB et al. VEGF expression in an osteoblast-like cell line is regulated by a hypoxia response mechanism. Am J Physiol Cell Physiol 2000; 278:C853-C860.
20. Ortega N, Behonick D, Stickens D et al. How proteases regulate bone morphogenesis. Ann NY Acad Sci 2003; 995:109-116.
21. Boyle WJ, Simonet WS, Lacey DL. Osteoclast differentiation and activation. Nature 2003; 423:337-342.
22. Barleon B, Sozzani S, Zhou D et al. Migration of human monocytes in response to vascular endothelial growth factor (VEGF) is mediated via the VEGF receptor VEGFR1. Blood 1996; 87:3336-3343.
23. Niida S, Kaku M, Amano H et al. Vascular endothelial growth factor can substitute for macrophage colony-stimulating factor in the support of osteoclastic bone resorption. J Exp Med 1999; 190:293-298.

24. Nakagawa M, Kaneda T, Arakawa T et al. Vascular endothelial growth factor (VEGF) directly enhances osteoclastic bone resorption and survival of mature osteoclasts. FEBS Lett 2000; 473:161-164.
25. Tombran-Tink J, Barnstable CJ. Osteoblasts and osteoclasts express PEDF, VEGF-A isoforms, and VEGF receptors: Possible mediators of angiogenesis and matrix remodeling in the bone. Biochem Biophys Res Commun 2004; 316:573-579.
26. Colnot C, Lu C, Hu D et al. Distinguishing the contributions of the perichondrium, cartilage, and vascular endothelium to skeletal development. Dev Biol 2004; 269:55-69.
27. Colnot CI, Helms JA. A molecular analysis of matrix remodeling and angiogenesis during long bone development. Mech Dev 2001; 100:245-250.
28. Blavier L, Delaissé JM. Matrix metalloproteinases are obligatory for the migration of preosteoclasts to the developing marrow cavity of primitive long bones. J Cell Sci 1995; 108:3649-3659.
29. Engsig MT, Chen QJ, Vu TH et al. Matrix metalloproteinase 9 and vascular endothelial growth factor are essential for osteoclast recruitment into developing long bones. J Cell Biol 2000; 151:879-889.
30. Henriksen K, Karsdal M, Delaisse JM et al. RANKL and vascular endothelial growth factor (VEGF) induce osteoclast chemotaxis through an ERK1/2-dependent mechanism. J Biol Chem 2003; 278:48745-48753.
31. Mayr-Wohlfart U, Waltenberger J, Hausser H et al. Vascular endothelial growth factor stimulates chemotactic migration of primary human osteoblasts. Bone 2002; 30:472-477.
32. Kitsukawa T, Shimono A, Kawakami A et al. Overexpression of a membrane protein, neuropilin, in chimeric mice causes anomalies in the cardiovascular system, nervous system and limbs. Development 1995; 121:4309-4318.
33. Vu TH, Shipley JM, Bergers G et al. MMP-9/gelatinase B is a key regulator of growth plate angiogenesis and apoptosis of hypertrophic chondrocytes. Cell 1998; 93:411-422.
34. Inada M, Wang Y, Byrne MH et al. Critical roles for collagenase-3 (Mmp13) in development of growth plate cartilage and in endochondral ossification. Proc Natl Acad Sci USA 2004; 101:17192-17197.
35. Stickens D, Behonick DJ, Ortega N et al. Altered endochondral bone development in matrix metalloproteinase 13-deficient mice. Development 2004; 131:5883-5895.
36. Ruhrberg C, Gerhardt H, Golding M et al. Spatially restricted patterning cues provided by heparin-binding VEGF-A control blood vessel branching morphogenesis. Genes Dev 2002; 16:2684-2698.
37. Midy V, Plouët J. Vasculotropin/vascular endothelial growth factor induces differentiation in cultured osteoblasts. Biochem Biophys Res Commun 1994; 199:380-386.
38. Hiltunen MO, Ruuskanen M, Huuskonen J et al. Adenovirus-mediated VEGF-A gene transfer induces bone formation in vivo. FASEB J 2003; 17:1147-1149.
39. Clauss M, Weich H, Breier G et al. The vascular endothelial growth factor receptor VEGFR1 mediates biological activities. Implications for a functional role of placenta growth factor in monocyte activation and chemotaxis. J Biol Chem 1996; 271:17629-17634.
40. Roach HI, Baker JE, Clarke NM. Initiation of the bony epiphysis in long bones: Chronology of interactions between the vascular system and the chondrocytes. J Bone Miner Res 1998; 13:950-961.
41. Holmbeck K, Bianco P, Caterina J et al. MT1-MMP-deficient mice develop dwarfism, osteopenia, arthritis, and connective tissue disease due to inadequate collagen turnover. Cell 1999; 99:81-92.
42. Zhou Z, Apte SS, Soininen R et al. Impaired endochondral ossification and angiogenesis in mice deficient in membrane-type matrix metalloproteinase I. Proc Natl Acad Sci USA 2000; 97:4052-4057.
43. Sounni NE, Roghi C, Chabottaux V et al. Up-regulation of vascular endothelial growth factor-A by active membrane-type 1 matrix metalloproteinase through activation of Src-tyrosine kinases. J Biol Chem 2004; 279:13564-13574.
44. Pfander D, Cramer T, Schipani E et al. HIF-1alpha controls extracellular matrix synthesis by epiphyseal chondrocytes. J Cell Sci 2003; 116:1819-1826.
45. Pfander D, Kobayashi T, Knight MC et al. Deletion of Vhlh in chondrocytes reduces cell proliferation and increases matrix deposition during growth plate development. Development 2004; 131:2497-2508.
46. Stalmans I, Lambrechts D, De Smet F et al. VEGF: A modifier of the del22q11 (DiGeorge) syndrome? Nat Med 2003; 9:173-182.
47. Lambrechts D, Storkebaum E, Morimoto M et al. VEGF is a modifier of amyotrophic lateral sclerosis in mice and humans and protects motoneurons against ischemic death. Nat Genet 2003; 34:383-394.

CHAPTER 8

VEGF in the Nervous System

Jeffrey M. Rosenstein, Janette M. Krum and Christiana Ruhrberg*

Abstract

Vascular endothelial growth factor (VEGF, VEGFA) is critical for blood vessel growth in the developing and adult nervous system of vertebrates. Several recent studies demonstrate that VEGF also promotes neurogenesis, neuronal patterning, neuroprotection and glial growth. For example, VEGF treatment of cultured neurons enhances survival and neurite growth independently of blood vessels. Moreover, evidence is emerging that VEGF guides neuronal migration in the embryonic brain and supports axonal and arterial copatterning in the developing skin. Even though further work is needed to understand the various roles of VEGF in the nervous system and to distinguish direct neuronal effects from indirect, vessel-mediated effects, VEGF can be considered a promising tool to promote neuronal health and nerve repair.

Key Messages
- VEGF promotes neurogenesis.
- VEGF has trophic effects on neurons and glia in the CNS and PNS.
- VEGF supports neuronal migration in the developing CNS.
- VEGF is essential for neuroprotection in adults.
- VEGF may be of therapeutic value in the treatment of neural disorders.

Introduction

The cytokine vascular endothelial growth factor (VEGF) fulfils several critical functions in blood vessels, both in adult vertebrates and during their development.[1,2] Accordingly, VEGF plays a central role in wound healing, tumour angiogenesis and retinopathies,[3-5] and loss of VEGF or its tyrosine kinase receptors causes severe vascular defects and therefore early embryonic demise in the mouse.[6-10] The role of VEGF in brain and retinal angiogenesis has been studied extensively. The most topical issue in VEGF biology, however, is the concept that VEGF has significant nonvascular functions in the nervous system. In this book chapter, we will discuss recent data supporting the idea of neurotrophic and instructive roles for VEGF in the nervous system that extend beyond its vascular functions to complement several previous reviews (refs. 11-13).

VEGF Isoforms and VEGF Receptors in the Nervous System

VEGF is essential for mouse development, as it promotes the condensation of endothelial cells into blood vessel networks in a process termed vasculogenesis.[8,9] Importantly, VEGF is normally made as a collection of three major isoforms that are produced by alternative splicing

*Corresponding Author: Christiana Ruhrberg—Institute of Ophthalmology, University College London, 11-43 Bath Street, London EC1V 9EL, UK. Email: c.ruhrberg@ucl.ac.uk

VEGF in Development, edited by Christiana Ruhrberg. ©2008 Landes Bioscience and Springer Science+Business Media.

and are coexpressed in tissue-specific ratios (see chapter 1 by Y. S. Ng).[1] These isoforms consist of 121, 165, or 189 amino acids in humans and 120, 164 and 188 amino acids in mice. The VEGF188 isoform is retained in the extracellular matrix after secretion, due to its high affinity for heparan sulphate proteoglycans (HSPGs). VEGF120 and VEGF164 are diffusible, although VEGF164 is also able to bind HSPGs. Amazingly, retention of any one of the VEGF isoforms rescues many vascular defects of full VEGF knockouts, presumably because each isoform is sufficient to support endothelial cell formation and proliferation. In contrast, the VEGF isoforms play different roles during later stages of vascular development. Based on a differential affinity for HSPGs in the extracellular matrix, the isoforms cooperate to establish chemoattractive gradients around VEGF-secreting cells and attract blood vessel sprouts from preexisting blood vessels in a process termed angiogenesis[14] (see chapter 6 by H. Gerhardt). In this fashion, VEGF isoform expression supports the formation of microvessel networks with optimal density. In addition, VEGF isoforms display a differential affinity for VEGF receptors. The tyrosine kinases KDR (also known as VEGFR2 or FLK1) and FLT1 (VEGFR1) have a high affinity for all VEGF isoforms. In contrast, the nontyrosine kinase neuropilin receptors NRP1 and NRP2 bind VEGF165, and possibly VEGF189, but not VEGF121.[1]

VEGF Controls Brain Angiogenesis

VEGF-induced blood vessel growth is essential for nervous tissue growth during embryonic development. This is demonstrated by the observation that loss of VEGF expression by central nervous system (CNS) neurons impairs vascularisation, curbs neuronal expansion and results in neuronal apoptosis in the developing brain.[15,16] It was first hypothesized more than 20 years ago that VEGF is synthesized by rapidly growing neuronal precursors to form a chemoattractive gradient that recruits blood vessels from the perineural vascular plexus to the subventricular zone.[17,18] This idea was corroborated by the expression pattern of VEGF and its receptors[19] as well as genetic studies demonstrating the role of VEGF gradients in guiding sprouting vessels in the brain.[14] VEGF gradients also guide blood vessels during retinal angiogenesis.[20] In the adult, VEGF is upregulated following injury to the CNS, and the exogenous application of VEGF promotes CNS angiogenesis. VEGF also increases blood-brain barrier permeability and may play a role in CNS inflammation.[21-25] KDR is essential for the formation of blood vessels in the embryo, as it controls endothelial cell differentiation, endothelial cell assembly into vascular networks and blood vessel sprouting, whereas FLT1 is crucial for vascular development, because it modulates VEGF/KDR signalling[7,26,27] (see chapter 5 by J. J. Haigh and chapter 4 by L. C. Goldie, M. K. Nix and K. K. Hirschi). The requirement for KDR and FLT1 specifically in CNS vascularisation has not been examined due to the early lethality of full knockout mice. In contrast, NRP1 knockout mice survive long enough to form a multilayered brain and spinal cord, and it was demonstrated that this VEGF receptor is essential for vessel growth in the CNS.[10,28] Endothelial cells express NRP1, and loss of NRP1 from endothelial cells impairs the vascularisation of the brain and spinal cord.[29,30] The most popular hypothesis of NRP1 function in vessel growth suggests that VEGF164 binds to NRP1 on endothelial cells to potentiate KDR signalling.[31]

Neurotrophic Roles of VEGF during Development and in the Adult

The Neurovascular Connection

During development, impaired CNS vascularisation inhibits neuronal proliferation and survival and thereby decreases cortical thickness in mice.[15,16] Recent insights indicate that alterations or loss of vascular patterning factors also contribute to neurodegeneration in adults. For example, a reduced serum VEGF level, due to mutation of the hypoxia response element of the *Vegfa* promoter, renders mice unusually sensitive to transient spinal cord ischemia; they remain paralyzed after a minor ischemic insult, whereas wild-type mice show only a transient clinical deficit.[32] Moreover, mice with a reduced serum level of VEGF develop a condition akin to amyotrophic lateral sclerosis (ALS).[33] ALS is a devastating disease characterized by progressive paralysis due to motor neuron degeneration, and it inevitably causes death. More than 90% of ALS patients are previously healthy with no family history, and the only molecular risk factor known to date is a reduced serum level of VEGF, due to a promoter polymorphism in the *VEGF* gene.[32] Low levels of VEGF have also been found in another motor neuron disease, spinal bulbar muscular atrophy.[34] Excitingly, VEGF treatment halts ALS-like motor neuron degeneration in mice.[35] Endothelial cells are likely targets in this kind of VEGF therapy, but it is not yet known if other cell types in the nervous system also respond to the administered VEGF. This is an important question, because VEGF affects astrocytes, Schwann cells and motor neurons in cell culture models. In the following paragraphs, we will review evidence consistent with the idea that VEGF affects blood vessel endothelium, glia and neurons in neurodegenerative diseases.

VEGF May Affect Neurodegeneration by Acting on Blood Vessels

Mice with reduced VEGF levels due to the mutation of the hypoxia response element in the *Vegfa* promoter show reduced neural tissue perfusion.[33] This was originally thought to be caused by inefficient vessel growth in the CNS during development. However, these mice did not show reduced capillary densities in the CNS, even though VEGF levels were reduced by 25% in the CNS.[33] Alternatively, reduced neural perfusion may be due to impaired vasoregulation, as VEGF affects vascular tone by controlling the release of the vasorelaxant nitric oxide by endothelial cells; accordingly, it has been hypothesised that VEGF is required for the preservation or the function of perivascular autonomic nerves, which regulate vascular tone.[13] Perfusion deficits in other neurodegenerative disorders, including Alzheimer's disease and Huntington's disease, may precede the onset of clinical symptoms, suggesting that they also contribute to the pathogenesis of these disorders. However, whether low VEGF levels contribute to impaired neural perfusion in these disorders remains to be elucidated. Determining the precise role of VEGF in these sitations will, however, be complicated by the fact that hypoxia, resulting from reduced perfusion, would likely lead to a secondary upregulation of VEGF (see chapter 3 by M. Fruttiger).

Vessel-Independent Trophic Roles for VEGF in the Nervous System

In addition to its vascular roles in the nervous system, VEGF may be a trophic factor for neurons that acts independently of the blood circulation. In support of this idea, VEGF treatment enhances neuronal survival and neurite outgrowth in explanted brain cortex or substantia nigra.[36,37] VEGF application also induces neurite outgrowth and enhances neuronal survival in cultured dorsal root ganglia.[38,39] Moreover, VEGF protects cultured cerebral neurons or hippocampal neurons during hypoxia or serum withdrawal,[40,41] and it also protects cultured hippocampal neurons against glutamate or NMDA toxicity.[42,43] Finally, overexpression of KDR in motor neurons protects them from cell death by pathogenic SOD1 mutant protein in vivo.[35] It is not yet clear if these neurotrophic effects of VEGF reflect a direct effect of VEGF on neurons or if they are mediated indirectly by glial cells. For example, VEGF promotes the survival of primary motor neurons on glial feeder layers following oxidative stress, an effect that is inhibited by antibodies that neutralize VEGF receptor activity.[33] Moreover, one of the growth factors secreted by astrocytes and Schwann cells is VEGF itself. VEGF may therefore have a dual neurotrophic effect, an indirect effect that is mediated by glia and a direct effect on neurons independently of glia. Consistent with the idea that VEGF could be neurotrophic by acting through glia, VEGF is mitogenic for astroglia and Schwann cells,[22,37,38,44-46] and both glial cell types produce a number of growth factors that support neuronal growth in explant cultures.[47,48]

Our own experiments support the idea of a direct neurotrophic role of VEGF on neurons. We found that VEGF promoted neurite extension in spinal cord explants (Fig. 1; J. K. and J. R., unpublished observations). This organotypic explant system was used to study VEGF's effect on neurons in the context of three-dimensional tissue architecture, in order to maintain density-dependant regulatory mechanisms and tissue specific diffusion properties. However, as

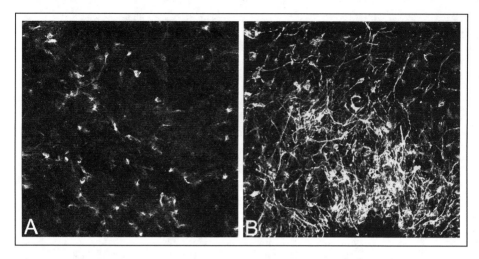

Figure 1. VEGF promotes neurite extension in spinal cord explants. Immunostaining of neurons in dorsal horn explants shows modest expression of MAP2 (A). When 50 ng/ml VEGF is added to lumbar spinal cord explants, MAP2 expression is upregulated in enlarged cell bodies as well as in elongated and branched neurites (B).

is the case when studying VEGF effects in the nervous system in vivo, this explant model did not distinguish direct effects of VEGF on neurons from indirect effects mediated by other VEGF-responsive cell types. For instance, VEGF application stimulated angiogenesis, which then supported astrocyte growth and thereby neuronal survival. That VEGF has a direct neurotrophic role for cultured neurons independently of glia was supported by the observation that VEGF treatment promoted neurite growth and maturation in cultures of primary cortical neurons lacking glia with similar efficiency to that in explant cultures containing glia[36,49] (Fig. 2). Furthermore, these primary cell culture studies suggested that VEGF upregulates neuron-specific enolase (NSE) and microtubule associated protein 2 (MAP2), possibly by stabilizing them or by up-regulating their expression.[11,36] MAP2 and several other proteins are downregulated after neuronal injury, and VEGF application might help to raise the levels of some of these proteins again to contribute to neuroprotection. In the future, tissue-specific genetic model systems will advance our understanding of VEGF's role on neurons, as they can distinguish VEGF's cell-type specific activities in an in vivo context.

VEGF Signalling in Neurons

KDR has been implicated as a VEGF receptor in mature neurons in several culture models,[33,39,50] and the alternative VEGF receptors FLT1, NRP1 and NRP2 have also been detected in several types of developing and adult neurons. For example, NRP1 is upregulated on neurons and endothelium after CNS ischemia,[51] suggesting roles in brain injury. However, the functional requirements for these different VEGF receptors in neurons were initally difficult to determine, owing to the severe cardiovascular defects and early lethality caused by loss of these proteins in the embryo (see above). Fortunately, several novel genetic tools using *Cre/Lox* technology have recently been developed to faciliate the creation of neuron-specific mutations in VEGF receptors, which circumvents the embryonic lethality caused by cardiovascular defects. Surprisingly, the disruption of the *Kdr* gene in CNS neurons with a nestin-based recombination approach did not impair neuronal development or viability in the mouse.[15] This finding suggested that KDR is not essential for neuronal growth and patterning. The genetic requirement for FLT1 as a VEGF receptor in neurons has not been examined so far, but we do know that FLT1's tyrosine kinase activity is dispensable for both embryogenesis

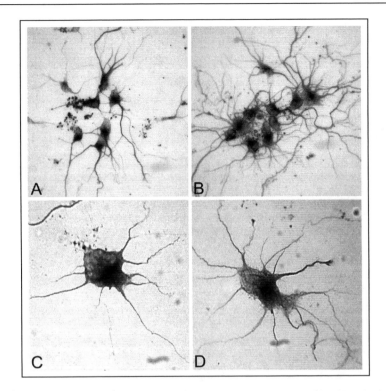

Figure 2. VEGF promotes neurite extension from cortical neurons. In primary cultures of cortical neurons (> 95% pure), neurofilament-positive neurites extend from neurons that are growing in clusters (A) or as single cells (C) in a serum-free environment. The addition of 50-100 ng/ml VEGF increases the diameter and length of neurites from clustered neurons significantly (B), with a stimulation of neurite outgrowth from the cell body by 30-40% (D).

and postnatal survival.[26] Therefore, any possible FLT1 function in neurons would be expected to involve its extracellular domain only. So far, it has been difficult to establish if neuropilins act as physiological VEGF receptors in neurons through loss of function studies, because they also transmit signals provided by neural guidance molecules of the class 3 semaphorin family, such as SEMA3A and SEMA3F.[52] In fact, we presently know of only one neuronal cell type in which NRP1 acts predominantly as a VEGF164 rather than a semaphorin receptor: the neuropilin-expressing, but SEMA3A-unresponsive facial branchiomotor neurons in the mouse brainstem (see below).

VEGF and Neurogenesis

In the embryonic nervous system, neuronal progenitors proliferate, migrate and differentiate in a process termed neurogenesis to populate the growing brain and ganglia with neurons. Embryonic neurogenesis originates from neuroepithelial progenitors in the subventricular zone of the CNS, and from neural crest cell-derived progenitors in the peripheral nervous system. In the adult, neurogenesis occurs mainly in two regions of the brain; (a) in the subventricular zone lining the lateral ventricles to produce the rostral migratory stream that supplies interneurons to the olfactory bulb; and (b) in the dentate gyrus of the hippocampus, where new neurons are thought to participate in memory formation. Neurogenesis can also be induced in the adult brain in response to pathological situations, such as mechanical trauma, seizures and ischemia. It has been hypothesized that neurogenesis in the adult brain relies on neural stem cells. However, the origin of neural stem cells in the adult brain remains to be defined. Surprisingly, they have characteristics of fully differentiated glia, and it has therefore been proposed that adult neuronal stem cells are derived from a cell lineage that initially forms embryonic neuroepithelial stem cells, then radial glia with the potential to form astrocytes and neurons, and finally astrocyte-like adult stem cells. Several different mitogenic and/or trophic growth factors have been implicated in the process of neurogenesis. Fibroblast growth factor 2 (FGF2, also known as basic FGF) and epidermal growth factor (EGF) in particular are mitogens for neural progenitor and stem cells in vitro. In addition, neurotrophic factors such as brain-derived neurotrophic factor (BDNF) are also involved in neurogenesis. VEGF expression by neurons is prominent in the developing brain and during brain pathology, when it may play a dual role to promote neurogenesis, firstly by acting as a paracrine factor for endothelial cells to stimulate pro-neurogenic angiogenesis (see above) and secondly, as an autocrine factor for neuronal progenitors.[50] Consistent with this idea, administration of VEGF to adult rat brain via an osmotic minipump stimulates neurogenesis, astrocyte production and endothelial cell growth in the hippocampus and the lateral subventricular zone.[53] We will first review evidence supporting a role for VEGF-stimulated angiogenesis in promoting neurogenesis and then discuss data consistent with a direct neurogenic role of VEGF.

VEGF and the Angiogenic Niche of Neurogenesis

Adult neurogenesis in the residual germinal matrices of the brain occurs in parallel with the growth of new blood vessels, which are thought to provide an "angiogenic niche" for the neuronal progenitors. This concept was first proposed to explain the correlation of neuronal progenitor mitoses and endothelial cell growth in the pseudo-glomerular structures of the dentate gyrus in the hippocampus.[54] Blood vessels in the angiogenic niche of neurogenesis could potentially provide instructive neurotrophic factors that promote the differentiation or survival of neuronal progenitors. This idea is supported by findings in the adult songbird brain, where neurogenesis proceeds throughout life in the higher vocal centre. Here, testosterone upregulates VEGF and QUEK1 (bird KDR) to induce angiogenesis and stimulate BDNF release from blood vessel endothelium. BDNF then supports neuronal differentiation and migration of neurons derived from the ventricular zone of the higher vocal cord centre.[55] Moreover, pigment epithelium-derived factor (PEDF) is produced by ependymal and endothelial cells to induce self-renewal of cultured neural stem cells.[56] Alternatively, or additionally, blood vessels in the angiogenic niche of neurogenesis might provide a substrate for migrating neuronal progenitors during their journey from their germinal zone to the site of their differentiation; this substrate might be the vessel surface or the vessel-associated extracellular matrix. Surprisingly, experiments have not yet been performed that address how the angiogenic niche contributes to neurogenesis in the developing CNS.

VEGF Signalling in Neuronal Progenitors during Development

Several findings support the idea that VEGF has direct effects on neuronal progenitors in the developing brain. VEGF164 stimulates the migration survival of a neuroectodermal progenitor cell line.[57] Moreover, VEGF application enhances the proliferation of embryonic cortical neuronal progenitors in vitro, and the blockade of KDR signalling prevents their VEGF-induced proliferation.[53] VEGF also affects the migration of primary neuronal progenitors: FGF stimulates the proliferation of neuronal progenitors derived from the newborn rat rostral subventricular zone concomitantly with increasing expression of KDR and FLT1, and the FGF2-stimulated neuronal progenitors become responsive to VEGF, which acts as a chemoattractant via KDR.[58] The role of VEGF and its receptors in neurogenesis has also been studied in the developing retina. VEGF and KDR are expressed in the inner retina prior to its vascularisation and also in the avascular outer retina, and several different experimental approaches have raised the possibility that VEGF signalling plays a role in the development of the neural retina. For example, VEGF treatment promotes neurogenesis in the avascular chick retina, whilst KDR inhibition with a small interfering RNA blocks proliferation; similarly, inhibition of receptor tyrosine kinases, including KDR, blocks development of the inner mouse retina, which would normally contain Mueller and retinal ganglion cells.[59-62] VEGF also stimulates photoreceptor development from retinal progenitors in vitro.[63]

VEGF Signalling in Neuronal Stem Cells and Adult Neuronal Progenitors

VEGF/KDR signalling affects the fate of neuronal stem cells derived from the embryonic brain.[64] On the one hand, VEGF promotes the survival of definitive neural stem cells, which form in the developing brain after 8.5 dpc and persist throughout adulthood. On the other hand, VEGF inhibits the survival of primitive neural stem cells, present in the embryonic brain up to 8.5 dpc. Interestingly, primitive neural stem cells would not normally be exposed to high VEGF levels until after 8.5 dpc, when VEGF is upregulated in the neural tube to attract blood vessels from the perineural vascular plexus. These observations raise the possibility that VEGF contributes to a developmental switch that affects neural stem cell behaviour. KDR expression has also been reported in the proliferative zones of the adult rodent brain,[50,53,65] although the specificity of the antibodies used was demonstrated in only one of these studies.[65] When VEGF is administered to the brain at low concentrations, it stimulates the proliferation of KDR-expressing cells in the ventricular zone in vivo independently of its effect on blood vessels.[65] VEGF application was also found to promote the survival of neural stem cells in vitro in a KDR-dependent fashion.[65] The physiological significance of VEGF as an autocrine or paracrine signal for neuronal stem cells is not yet understood, as it has not been possible to ablate VEGF expression in the CNS without simultaneously affecting blood vessels.[15,16] The specific deletion of VEGF receptors in the early neuronal lineage may therefore provide a more suitable approach to determine the contribution of VEGF signalling to neurogenesis. The ablation of KDR expression from neurons using nestin-promoter driven CRE-mediated recombination suggested that this VEGF receptor is not essential for embryonic neurogenesis.[15] However, the impact of KDR loss from CNS neurons on neonatal and adult neurogenesis or neuronal survival has not yet been studied, and FLT1 and NRP1 have not yet been ablated specifically in the neuronal lineage. Further work on the role of VEGF and its receptors in the neuronal lineage has become pressing, given that reduced VEGF levels were found to cause motor neuron degeneration.

VEGF Signalling in Neurovascular Patterning

Several recent studies have explored the idea that endothelial cells and neurons share signalling pathways to control their growth and behaviour. Accordingly, there is now evidence that axon guidance cues control blood vessel branching and, vice versa, that vascular patterning molecules can modify the migration of neurons and glia. For example, ephrin/EPH signals and netrins with their UNC and DCC receptors have been implicated both in neuronal and vascular patterning,[66] and NRP1 plays a dual role as an isoform-specific VEGF receptor on endothelial cells[31] and a neuronal cell surface receptor for semaphorins.[67] The observation that VEGF165 and the semaphorin SEMA3A compete for NRP1 binding in cell culture models[68,69] raised the possibility that shared transmembrane receptors serve to coordinate vascular and neuronal growth by integrating antagonistic signals.[66] Several recent studies have begun to test if semaphorins control vascular development, if VEGF controls neuronal development, and if ligand sharing by NRP1 controls neuronal and vascular copatterning in vivo. The outcome of these studies is discussed in the following paragraphs.

VEGF in Neuronal Patterning

VEGF promotes neurite extension and neurite maturation in cell culture models (Figs. 1 and 2),[36-39] and the VEGF164 isoform has been hypothesised to act as an axonal guidance cue by binding to NRP1 in competition with SEMA3A.[70] Unfortunately, in vivo evidence that VEGF acts as an axonal patterning factor in NRP1-expressing neurons is so far lacking: In developing dorsal root ganglia, VEGF has no effect on neurons, even though it controls endothelial cell growth,[71] and limb axons grow normally in the absence of VEGF164.[72] VEGF164 does, however, pattern neuronal migration within the developing CNS, as it controls the NRP1-dependent cell body migration of facial branchiomotor neurons within the mouse brainstem.[73] Strikingly, there is no competition of VEGF164 and SEMA3A during this process. Rather, both NRP1 ligands cooperate by regulating different aspects of neuronal behaviour, with VEGF164 signalling being necessary only for the correct pathfinding of facial branchiomotor somata within the brainstem, and SEMA3A being involved solely in the guidance of their axons in the second branchial arch. Importantly, VEGF164 appears to pattern facial branchiomotor neurons independently of blood vessels, as the endothelial-specific NRP1 knockout displays vascular, but not neuronal migration defects. Similarly to facial branchiomotor axons, sensory and motor axons in the limb require SEMA3A, but not VEGF164 to control their growth. Whilst the developmental analysis of facial branchiomotor neurons has provided the first evidence that VEGF can directly control neuronal behaviour independently of blood vessels, further work is required to examine if VEGF patterns other neuronal cell types and can therefore be considered a general neuronal patterning molecule like SEMA3A. In particular, it will be important to address if VEGF effects are restricted to the guidance of neuronal cell bodies, or if some types of axons also use VEGF as a migratory cue. Candidate neurons whose axons may use VEGF164 as a guidance molecule may be identified by their ability to express NRP1, even though they do not require SEMA3A signals for their patterning. The observation that VEGF promotes neurite maturation (see above) raises the possibility that VEGF might affect dendrite development.

A Role for Competition of VEGF164 and SEMA3A during Neuronal and Vascular Patterning?

Tissue culture studies have shown that VEGF165 and SEMA3A compete for binding to the extracellular domain of NRP1.[57,68] Moreover, in experiments with chick limbs carrying SEMA3A bead implants, vessels and nerves were both repelled by SEMA3A in a mechanism requiring NRP1.[74] Based on these observations, it has been hypothesised that VEGF165 and SEMA3A coordinate vascular and neural development by competing for NRP1 and thereby contribute to the emergence of neurovascular congruence. However, the role of SEMA3A signalling in vascular development is presently controversial. SEMA3A has been implicated in vascular patterning in zebrafish, where one of two SEMA3A forms termed SEMA3AB (SEMA3A2) curbs intersegmental vessel branching.[75] In addition, Serini and coworkers reported that loss of SEMA3A impairs head vessel remodelling, intersomitic vessel branching and formation of the anterior cardinal vein by modulating integrin signalling in the mouse and chick.[76] In contrast, we and others have found that SEMA3A and semaphorin-signalling through NRP1 are not required for microvessel patterning during mouse development.[72,77] Moreover, we have found that there is no genetic interaction between VEGF164 and SEMA3A during vasculogenesis or angiogenic vessel growth and branching.[72] The observation that there is no competition between VEGF164 and SEMA3A during vessel growth in vivo is consistent with the finding that VEGF164, but not SEMA3A controls the cell body migration of facial branchiomotor neurons, whilst SEMA3A, but not VEGF164 guides facial nerve and limb axons (see above). The emerging picture is therefore one of ligand cooperation rather than competition, with VEGF164 and SEMA3A being specialised to mediate distinct patterning events that occur in close spatiotemporal proximity.

VEGF Mediates the Copatterning of Nerves and Arteries

Whilst ligand competition between VEGF164 and SEMA3A does not appear to be critical for neurovascular development, VEGF does nevertheless play a role in neurovascular copatterning independently of its relationship to SEMA3A: In the developing limb skin, VEGF is required for the alignment of nerves and arteries, because nerve-secreted VEGF acts on vascular NRP1 to promote arteriogenesis.[78,79] Notably, VEGF is likely to cooperate with other neuronal-derived signals in the copatterning of vessels and nerves in the limb, as it induces arterial character rather than mediating the recruitment of blood vessels to nerves. The factors that control the coalignment of nerves and vessels have therefore remained elusive.

VEGF and Nervous System Repair

VEGF production by CNS neurons early on in development is likely induced by hypoxia to match vessel growth to oxygen requirements and metabolic demand (see chapter 3 by M. Fruttiger). In an analogous fashion, VEGF may be induced during nervous system damage and peripheral nerve regeneration to stimulate new vessel growth and enhance tissue repair. Consistent with this hypothesis, several studies suggest that VEGF treatment improves diabetic and peripheral neuropathy. Firstly, the intramuscular gene transfer of a plasmid encoding VEGF enhances motor and sensory functions in a rabbit model of ischemic peripheral neuropathy.[45] Secondly, application of VEGF after stroke injury can decrease brain infarct size, likely by promoting angiogenesis and neurogenesis near the penumbral area.[23,25,80,81] Thirdly, application of function-blocking VEGF-specific antibodies in a stab injury model increased lesion size and decreased angiogenic and astroglial activity in the striatum.[82] In further support of a role for VEGF in damage control during brain injuries, application of VEGF to the contused spinal cord produced behavioural and cellular improvements,[83,84] and VEGF significantly enhanced nerve regeneration when applied to matrigel implants into injured sciatic nerves.[85]

Conclusions and Future Perspectives

The studies described in this review demonstrate that VEGF plays multiple roles in the nervous system by acting on blood vessels, glia and neurons; a working model for the role of VEGF in peripheral nerves is presented in Figure 3. Because of its multiple effects, VEGF

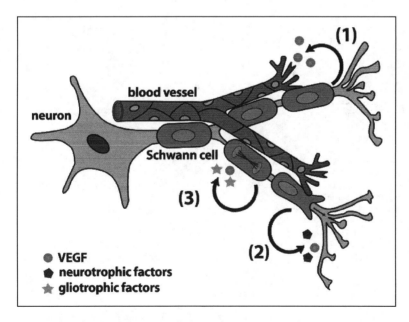

Figure 3. Possible roles for Schwann cell-derived VEGF in peripheral nerves. Schwann cells secrete VEGF, but also several other gliotrophic and neurotrophic factors. VEGF stimulates the growth and patterning of blood vessels (1) and is likely to help protect motor and sensory neurons (2). In addition, VEGF is thought to protect Schwann cells and stimulate their proliferation and migration in an autocrine loop (3). It is not yet known if paracrine VEGF sources for Schwann cells exist in peripheral nerves.

treatment may be beneficial for neurodegenerative and neuropathic conditions by enhancing both blood vessel and glial cell growth, whilst also providing direct neuroprotection. Whereas VEGF application after brain or spinal cord injury or in some neurodegenerative diseases is likely to aid neural and glial protection, VEGF may also have negative effects, as it can promote excessive macrophage infiltration, increase vascular permeability and disrupt the blood-brain barrier.[22-24,82,86] Whether VEGF treatment will ultimately be useful in the clinic to promote neoangiogenesis and neuronal survival may therefore depend on our ability to separate its effects on macrophages and vascular permeability from its angiogenic and neurotrophic effects. It will also be imperative to investigate if the use of anti-VEGF therapy to treat unwanted angiogenesis and vascular leakage in cancer and eye diseases could cause serious side effects in nervous tissue, as reduced VEGF levels may impair adult neurogenesis and neuroprotection. Investigating the contribution of VEGF and its signalling pathways to brain development is likely to provide critical clues that will help us understand VEGF's physiological functions and therefore help to pioneer novel therapeutic strategies for nervous system repair.

Acknowledgements

We thank Drs. N. Mani and N. More for their expertise in tissue culture work and Dr Q. Schwarz for (Fig. 3). J. M. K. and J. M. R. are supported by NS-39282 and NS-45189. C. R. is supported by an MRC Career Development Award.

References

1. Ruhrberg C. Growing and shaping the vascular tree: Multiple roles for VEGF. Bioessays 2003; 25(11):1052-1060.
2. Ferrara N, Davis-Smyth T. The biology of vascular endothelial growth factor. Endocr Rev 1997; 18(1):4-25.
3. Dvorak HF, Brown LF, Detmar M et al. Vascular permeability factor/vascular endothelial growth factor, microvascular hyperpermeability, and angiogenesis. Am J Pathol 1995; 146(5):1029-1039.
4. Ferrara N. VEGF: An update on biological and therapeutic aspects. Curr Opin Biotechnol 2000; 11(6):617-624.
5. Ferrara N, Gerber HP, LeCouter J. The biology of VEGF and its receptors. Nat Med 2003; 9(6):669-676.
6. Fong GH, Rossant J, Gertsenstein M et al. Role of the Flt-1 receptor tyrosine kinase in regulating the assembly of vascular endothelium. Nature 1995; 376(6535):66-70.
7. Shalaby F, Rossant J, Yamaguchi TP et al. Failure of blood-island formation and vasculogenesis in Flk-1-deficient mice. Nature 1995; 376(6535):62-66.
8. Carmeliet P, Ferreira V, Breier G et al. Abnormal blood vessel development and lethality in embryos lacking a single VEGF allele. Nature 1996; 380(6573):435-439.
9. Ferrara N, Carver-Moore K, Chen H et al. Heterozygous embryonic lethality induced by targeted inactivation of the VEGF gene. Nature Apr 4 1996; 380(6573):439-442.
10. Kawasaki T, Kitsukawa T, Bekku Y et al. A requirement for neuropilin-1 in embryonic vessel formation. Development 1999; 126(21):4895-4902.
11. Rosenstein JM, Krum JM. New roles for VEGF in nervous tissue-beyond blood vessels. Exp Neurol 2004; 187(2):246-253.
12. Carmeliet P, Storkebaum E. Vascular and neuronal effects of VEGF in the nervous system: Implications for neurological disorders. Semin Cell Dev Biol 2002; 13(1):39-53.
13. Storkebaum E, Carmeliet P. VEGF: A critical player in neurodegeneration. J Clin Invest 2004; 113(1):14-18.
14. Ruhrberg C, Gerhardt H, Golding M et al. Spatially restricted patterning cues provided by heparin-binding VEGF-A control blood vessel branching morphogenesis. Genes Dev 2002; 16(20):2684-2698.
15. Haigh JJ, Morelli PI, Gerhardt H et al. Cortical and retinal defects caused by dosage-dependent reductions in VEGF-A paracrine signaling. Dev Biol 2003; 262(2):225-241.
16. Raab S, Beck H, Gaumann A et al. Impaired brain angiogenesis and neuronal apoptosis induced by conditional homozygous inactivation of vascular endothelial growth factor. Thromb Haemost 2004; 91(3):595-605.
17. Bar T. Patterns of vascularization in the developing cerebral cortex. Ciba Found Symp 1983; 100:20-36.
18. Risau W. Mechanisms of angiogenesis. Nature 1997; 386(6626):671-674.

19. Breier G, Albrecht U, Sterrer S et al. Expression of vascular endothelial growth factor during embryonic angiogenesis and endothelial cell differentiation. Development 1992; 114:521-532.
20. Gerhardt H, Golding M, Fruttiger M et al. VEGF guides angiogenic sprouting utilizing endothelial tip cell filopodia. J Cell Biol 2003; 161(6):1163-1177.
21. Rosenstein JM, Mani N, Silverman WF et al. Patterns of brain angiogenesis after vascular endothelial growth factor administration in vitro and in vivo. Proc Natl Acad Sci USA 1998; 95(12):7086-7091.
22. Krum JM, Mani N, Rosenstein JM. Angiogenic and astroglial responses to vascular endothelial growth factor administration in adult rat brain. Neuroscience 2002; 110(4):589-604.
23. Zhang ZG, Zhang L, Jiang Q et al. VEGF enhances angiogenesis and promotes blood-brain barrier leakage in the ischemic brain. J Clin Invest 2000; 106(7):829-838.
24. Proescholdt MA, Heiss JD, Walbridge S et al. Vascular endothelial growth factor (VEGF) modulates vascular permeability and inflammation in rat brain. J Neuropathol Exp Neurol 1999; 58(6):613-627.
25. van Bruggen N, Thibodeaux H, Palmer JT et al. VEGF antagonism reduces edema formation and tissue damage after ischemia/reperfusion injury in the mouse brain. J Clin Invest 1999; 104(11):1613-1620.
26. Hiratsuka S, Minowa O, Kuno J et al. Flt-1 lacking the tyrosine kinase domain is sufficient for normal development and angiogenesis in mice. Proc Natl Acad Sci USA 1998; 95(16):9349-9354.
27. Kearney JB, Kappas NC, Ellerstrom C et al. The VEGF receptor flt-1 (VEGFR-1) is a positive modulator of vascular sprout formation and branching morphogenesis. Blood 2004; 103(12):4527-4535.
28. Gerhardt H, Ruhrberg C, Abramsson A et al. Neuropilin-1 is required for endothelial tip cell guidance in the developing central nervous system. Dev Dyn 2004; 231(3):503-509.
29. Kitsukawa T, Shimono A, Kawakami A et al. Overexpression of a membrane protein, neuropilin, in chimeric mice causes anomalies in the cardiovascular system, nervous system and limbs. Development 1995; 121(12):4309-4318.
30. Gu C, Rodriguez ER, Reimert DV et al. Neuropilin-1 conveys semaphorin and VEGF signaling during neural and cardiovascular development. Dev Cell 2003; 5(1):45-57.
31. Soker S, Takashima S, Miao HQ et al. Neuropilin-1 is expressed by endothelial and tumor cells as an isoform-specific receptor for vascular endothelial growth factor. Cell 1998; 92(6):735-745.
32. Lambrechts D, Storkebaum E, Morimoto M et al. VEGF is a modifier of amyotrophic lateral sclerosis in mice and humans and protects motoneurons against ischemic death. Nat Gen 2003; 34(4):383-393.
33. Oosthuyse B, Moons L, Storkebaum E et al. Deletion of the hypoxia-response element in the vascular endothelial growth factor promoter causes motor neuron degeneration.[comment]. Nat Gene 2001; 28(2):131-138.
34. Sopher BL, Thomas Jr PS, LaFevre-Bernt MA et al. Androgen receptor YAC transgenic mice recapitulate SBMA motor neuronopathy and implicate VEGF164 in the motor neuron degeneration. Neuron 2004; 41(5):687-699.
35. Storkebaum E, Lambrechts D, Dewerchin M et al. Treatment of motoneuron degeneration by intracerebroventricular delivery of VEGF in a rat model of ALS. Nat Neurosci 2005; 8(1):85-92.
36. Rosenstein JM, Mani N, Khaibullina A et al. Neurotrophic effects of vascular endothelial growth factor on organotypic cortical explants and primary cortical neurons. J Neurosci 2003; 23(35):11036-11044.
37. Silverman WF, Krum JM, Mani N et al. Vascular, glial and neuronal effects of vascular endothelial growth factor in mesencephalic explant cultures. Neuroscience 1999; 90(4):1529-1541.
38. Sondell M, Lundborg G, Kanje M. Vascular endothelial growth factor has neurotrophic activity and stimulates axonal outgrowth, enhancing cell survival and Schwann cell proliferation in the peripheral nervous system. Neuroscience 1999; 19(14):5731-5740.
39. Sondell M, Sundler F, Kanje M. Vascular endothelial growth factor is a neurotrophic factor which stimulates axonal outgrowth through the flk-1 receptor. Eur J Neurosci 2000; 12(12):4243-4254.
40. Jin KL, Mao XO, Nagayama T et al. Induction of vascular endothelial growth factor and hypoxia-inducible factor-1alpha by global ischemia in rat brain. Neuroscience 2000; 99(3):577-585.
41. Jin KL, Mao XO, Greenberg DA. Vascular endothelial growth factor: Direct neuroprotective effect in in vitro ischemia. Proc Natl Acad Sci USA 2000; 97(18):10242-10247.
42. Matsuzaki H, Tamatani M, Yamaguchi A et al. Vascular endothelial growth factor rescues hippocampal neurons from glutamate-induced toxicity: Signal transduction cascades. FASEB J 2001; 15(7):1218-1220.

43. Svensson B, Peters M, Konig HG et al. Vascular endothelial growth factor protects cultured rat hippocampal neurons against hypoxic injury via an antiexcitotoxic, caspase-independent mechanism. J Cereb Blood Flow Metab 2002; 22(10):1170-1175.
44. Sondell M, Lundborg G, Kanje M. Vascular endothelial growth factor stimulates Schwann cell invasion and neovascularization of acellular nerve grafts. Brain Res 1999; 846(2):219-228.
45. Schratzberger P, Schratzberger G, Silver M et al. Favorable effect of VEGF gene transfer on ischemic peripheral neuropathy. Nat Med 2000; 6(4):405-413.
46. Mani N, Khaibullina A, Krum JM et al. Astrocyte growth effects of vascular endothelial growth factor (VEGF) application to perinatal neocortical explants: Receptor mediation and signal transduction pathways. Exp Neurol 2005; 192:394-406.
47. Eddleston M, Mucke L. Molecular profile of reactive astrocytes-implications for their role in neurologic disease. Neuroscience 1993; 54(1):15-36.
48. Krum JM, Rosenstein JM. VEGF mRNA and its receptor flt-1 are expressed in reactive astrocytes following neural grafting and tumor cell implantation in the adult CNS. Exp Neurol 1998; 154(1):57-65.
49. Khaibullina AA, Rosenstein JM, Krum JM. Vascular endothelial growth factor promotes neurite maturation in primary CNS neuronal cultures. Brain Research. Dev Brain Res 2004; 148(1):59-68.
50. Ogunshola OO, Antic A, Donoghue MJ et al. Paracrine and autocrine functions of neuronal vascular endothelial growth factor (VEGF) in the central nervous system. J Biol Chem 2002; 277(13):11410-11415.
51. Zhang ZG, Tsang W, Zhang L et al. Up-regulation of neuropilin-1 in neovasculature after focal cerebral ischemia in the adult rat. J Cereb Blood Flow Metab 2001; 21(5):541-549.
52. Raper JA. Semaphorins and their receptors in vertebrates and invertebrates. Curr Opin Neurobiol 2000; 10(1):88-94.
53. Jin K, Zhu Y, Sun Y et al. Vascular endothelial growth factor (VEGF) stimulates neurogenesis in vitro and in vivo. Proc Natl Acad Sci USA 2002; 99(18):11946-11950.
54. Palmer TD, Willhoite AR, Gage FH. Vascular niche for adult hippocampal neurogenesis. J Comp Neurol 2000; 425(4):479-494.
55. Louissaint Jr A, Rao S, Leventhal C et al. Coordinated interaction of neurogenesis and angiogenesis in the adult songbird brain. Neuron 2002; 34(6):945-960.
56. Ramirez-Castillejo C, Sanchez-Sanchez F, Andreu-Agullo C et al. Pigment epithelium-derived factor is a niche signal for neural stem cell renewal. Nat Neurosci 2006; 9(3):331-339.
57. Bagnard D, Vaillant C, Khuth ST et al. Semaphorin 3A-vascular endothelial growth factor-165 balance mediates migration and apoptosis of neural progenitor cells by the recruitment of shared receptor. J Neurosci 2001; 21(10):3332-3341.
58. Zhang H, Vutskits L, Pepper MS et al. VEGF is a chemoattractant for FGF-2-stimulated neural progenitors. J Cell Biol 2003; 163(6):1375-1384.
59. Yang K, Cepko CL. Flk-1, a receptor for vascular endothelial growth factor (VEGF), is expressed by retinal progenitor cells. J Neurosci 1996; 16(19):6089-6099.
60. Robinson GS, Ju M, Shih SC et al. Nonvascular role for VEGF: VEGFR-1, 2 activity is critical for neural retinal development. FASEB J 2001; 15(7):1215-1217.
61. Gariano RF, Hu D, Helms J. Expression of angiogenesis-related genes during retinal development. Gene Expr Patterns 2006; 6(2):187-192.
62. Hashimoto T, Zhang XM, Chen BY et al. VEGF activates divergent intracellular signaling components to regulate retinal progenitor cell proliferation and neuronal differentiation. Development 2006; 133(11):2201-2210.
63. Yourey PA, Gohari S, Su JL et al. Vascular endothelial cell growth factors promote the in vitro development of rat photoreceptor cells. J Neurosci 2000; 20(18):6781-6788.
64. Wada T, Haigh JJ, Ema M et al. Vascular endothelial growth factor directly inhibits primitive neural stem cell survival but promotes definitive neural stem cell survival. J Neurosci 2006; 26(25):6803-6812.
65. Schanzer A, Wachs FP, Wilhelm D et al. Direct stimulation of adult neural stem cells in vitro and neurogenesis in vivo by vascular endothelial growth factor. Brain Pathol 2004; 14(3):237-248.
66. Eichmann A, Makinen T, Alitalo K. Neural guidance molecules regulate vascular remodeling and vessel navigation. Genes Dev 2005; 19(9):1013-1021.
67. Kolodkin AL, Ginty DD. Steering clear of semaphorins: Neuropilins sound the retreat. Neuron 1997; 19(6):1159-1162.
68. Miao HQ, Soker S, Feiner L et al. Neuropilin-1 mediates collapsin-1/semaphorin III inhibition of endothelial cell motility: Functional competition of collapsin-1 and vascular endothelial growth factor-165. J Cell Biol 1999; 146(1):233-242.

69. Bagnard D, Lohrum M, Uziel D et al. Semaphorins act as attractive and repulsive guidance signals during the development of cortical projections. Development 1998; 125(24):5043-5053.
70. Carmeliet P. Blood vessels and nerves: Common signals, pathways and diseases. Nat Rev Genet 2003; 4(9):710-720.
71. Kutcher ME, Klagsbrun M, Mamluk R. VEGF is required for the maintenance of dorsal root ganglia blood vessels but not neurons during development. FASEB J 2004; 18(15):1952-1954.
72. Vieira JM, Schwarz Q, Ruhrberg C. Selective requirements for neuropilin lignads in neurovascular development. Development 2007, (in press).
73. Schwarz Q, Gu C, Fujisawa H et al. Vascular endothelial growth factor controls neuronal migration and cooperates with Sema3A to pattern distinct compartments of the facial nerve. Genes Dev 2004; 18(22):2822-2834.
74. Bates D, Taylor GI, Minichiello J et al. Neurovascular congruence results from a shared patterning mechanism that utilizes Semaphorin3A and Neuropilin-1. Dev Biol 2003; 255(1):77-98.
75. Torres-Vazquez J, Gitler AD, Fraser SD et al. Semaphorin-plexin signaling guides patterning of the developing vasculature. Dev Cell 2004; 7(1):117-123.
76. Serini G, Valdembri D, Zanivan S et al. Class 3 semaphorins control vascular morphogenesis by inhibiting integrin function. Nature 2003; 424(6947):391-397.
77. Gu C, Yoshida Y, Livet J et al. Semaphorin 3E and plexin-D1 control vascular pattern independently of neuropilins. Science 2005; 307(5707):265-268.
78. Mukouyama YS, Gerber HP, Ferrara N et al. Peripheral nerve-derived VEGF promotes arterial differentiation via neuropilin 1-mediated positive feedback. Development 2005; 132(5):941-952.
79. Mukouyama YS, Shin D, Britsch S et al. Sensory nerves determine the pattern of arterial differentiation and blood vessel branching in the skin. Cell 2002; 109(6):693-705.
80. Sun Y, Jin K, Xie L et al. VEGF-induced neuroprotection, neurogenesis, and angiogenesis after focal cerebral ischemia. J Clin Invest 2003; 111(12):1843-1851.
81. Hayashi T, Abe K, Itoyama Y. Reduction of ischemic damage by application of vascular endothelial growth factor in rat brain after transient ischemia. J Cereb Blood Flow Metab 1998; 18(8):887-895.
82. Krum JM, Khaibullina A. Inhibition of endogenous VEGF impedes revascularization and astroglial proliferation: Roles for VEGF in brain repair. Exp Neurol 2003; 181(2):241-257.
83. Facchiano F, Fernandez E, Mancarella S et al. Promotion of regeneration of corticospinal tract axons in rats with recombinant vascular endothelial growth factor alone and combined with adenovirus coding for this factor. J Neurosurg 2002; 97(1):161-168.
84. Widenfalk J, Lipson A, Jubran M et al. Vascular endothelial growth factor improves functional outcome and decreases secondarydegeneration in experimental spinal cord contusion injury. Neuroscience 2003; 120:951-960.
85. Hobson MI, Green CJ, Terenghi G. VEGF enhances intraneural angiogenesis and improves nerve regeneration after axotomy. J Anatomy 2000; 197(Pt 4):591-605.
86. Kalaria RN, Cohen DL, Premkumar DR et al. Vascular endothelial growth factor in Alzheimer's disease and experimental cerebral ischemia. Brain Res Mol Brain Res 1998; 62(1):101-105.

Index